Highland Park Library
000201802305

HOW THE INTERNET HAPPENED

HOW THE INTERNET HAPPENED

FROM NETSCAPE TO THE IPHONE

BRIAN McCULLOUGH

LIVERIGHT PUBLISHING CORPORATION
A Division of W. W. NORTON & COMPANY
Independent Publishers Since 1923
NEW YORK LONDON

Printed in the United States of America
First Edition

For information about special discounts for bulk purchases, please contact W. W. Norton Special Sales at specialsales@wwnorton.com or 800-233-4830

Manufacturing by Worzalla
Book design by Barbara Bachman
Production manager: Julia Druskin

ISBN 978-1-63149-307-2

Liveright Publishing Corporation, 500 Fifth Avenue, New York, N.Y. 10110
www.wwnorton.com

W. W. Norton & Company Ltd., 15 Carlisle Street, London W1D 3BS

1 2 3 4 5 6 7 8 9 0

For:

My Dad, who taught me to love history

And for Werner Stocker, who taught me that history was cool

"Then there is electricity!—the demon, the angel, the mighty physical power, the all-pervading intelligence!" exclaimed Clifford. "Is that a humbug, too? Is it a fact—or have I dreamt it—that, by means of electricity, the world of matter has become a great nerve, vibrating thousands of miles in a breathless point of time? Rather, the round globe is a vast head, a brain, instinct with intelligence! Or, shall we say, it is itself a thought, nothing but thought, and no longer the substance which we deemed it!"

—*THE HOUSE OF THE SEVEN GABLES*,
CHAPTER 17,
"THE FLIGHT OF TWO OWLS,"
NATHANIEL HAWTHORNE

■

I think one of the things that really separates us from the high primates is that we're tool builders. I read a study that measured the efficiency of locomotion for various species on the planet. The condor used the least energy to move a kilometer. And, humans came in with a rather unimpressive showing, about a third of the way down the list. It was not too proud a showing for the crown of creation. So, that didn't look so good. But, then somebody at *Scientific American* had the insight to test the efficiency of locomotion for a man on a bicycle. And, a man on a bicycle, a human on a bicycle, blew the condor away, completely off the top of the charts.

And that's what a computer is to me. What a computer is to me is it's the most remarkable tool that we've ever come up with, and it's the equivalent of a bicycle for our minds.

—STEVE JOBS
1990 INTERVIEW FOR THE FILM
MEMORY & IMAGINATION

CONTENTS

HOW THE INTERNET HAPPENED

INTRO

When computers were first developed in the 1940s and '50s, it was never imagined that the common man or woman would ever need, much less have use for, them. Computers were designed for big problems: calculating missile trajectories; putting a man on the moon. Legend holds that the founder of IBM, Thomas J. Watson, once remarked, "I think there is a world market for maybe five computers." This quote is probably apocryphal, but it does capture the early thinking when computers first began to serve man. They were to be rare, expensive oracles; and like the oracles of ancient times, they would be useful only in rare, exceptional cases.

Computers were expensive in the beginning. They were complicated and difficult and they were as big as a room (that is not hyperbolic phrasing; generally considered the first modern computer, the ENIAC occupied about 1,800 square feet and weighed almost 50 tons). The received wisdom of their rarefied utility affected their design; computers were not conceived to be user-friendly because it was never assumed a nonexpert user would interact with one. Histories of early computers talk about a "priesthood" of computer specialists who actually interacted with the machines. Say you had a mathematical or engineering problem to solve. You submitted your punch cards to the "priests," and they used the computer to tease out the answer. Even when computers began to infiltrate the workplace in the 1960s through the 1980s (much to the surprise of the computer industry itself), it was still assumed that an "average" computer user could only achieve competency in limited, specific tasks or programs. The greater wrangling or mastery of the system at large was left to the early progenitors of what would become known as "the IT guy."

And yet, the tantalizing, almost forbidden mystique of computers seduced a generation of what were considered hobbyists in the 1970s. The hobbyists wanted to master computers themselves. They wanted computers that responded to them directly, without intermediaries. They wanted *personal* computing. And so, they made it happen. Steve Jobs, Steve Wozniak, Bill Gates, the Homebrew Computer Club—the hobbyists created the personal computer category (originally, they were called microcomputers) and thus, the PC Industry.

That still wasn't quite enough to make computers friendly to the average person. Almost a decade into the PC era, the industry remained trapped in the paradigm of the "command line." If you sat in front of a computer, you would see a blinking cursor and would need to type something to make the machine do anything for you. What would you need to type? Well, see, that's the point: it was functional inscrutability that continued to make computers so abstruse. In the era of the command line, you almost needed to have read the manual cover-to-cover or have previously mastered a computer language even to use the damn things. You had to know how to use a computer before you could use a computer.

This problem was solved by the invention of the GUI, or graphical user interface. Computers were humanized by graphics, by colors, by friendly icons and drop-down menus and a cute little tool called a mouse. Now when you sat in front of a computer, you could grab the mouse and just—click. You didn't have to know anything beforehand. You could learn how to use the machine by using it. Invented by Xerox, popularized by Apple Computers and the Macintosh, and then mainstreamed by Microsoft and its Windows operating system, the GUI was the evolutionary leap that would eventually make computers friendly to the average user.

But even when computers began to enter our everyday lives, our offices, our homes, they still were a bit esoteric. You might use a word processor at your job. Your kids might play computer games in your basement. But you didn't really need or use a computer in your everyday life. By 1990, only 42% of U.S. adults said they used a computer even "rarely."[1] In that same year, the number of American households that owned a computer had not yet passed 20%.[2]

■

THE INTERNET, AND ESPECIALLY the World Wide Web, finally brought computers into the mainstream. The Internet is *the* reason that computers actually became useful for the average person. The Internet is the thing that made a computer something you check in with daily, even hourly. And that is what this book is about: how the web and the Internet allowed computers to infiltrate our everyday lives. This is not a history of the Internet itself, but rather, a history of the Internet *Era*, that period of time from roughly 1993 through 2008 when computers and technology itself stopped being esoteric and started becoming vital and indispensable. It is about great technologies and disruptions and entrepreneurs. It is about how we allowed these technologies into our lives, and how these technologies changed us.

■

LIKE COMPUTERS, THE INTERNET was not designed with you and me in mind.

Computers were first hooked together in a meaningful way in 1969. This was the ARPANET, the grandfather of the Internet, and (mostly) true to legend, it was birthed by a Cold War–era alliance of the United States military and the academic-industrial complex. Funded by DARPA, the Defense Advanced Research Projects Agency, the Internet's first four connections, or "nodes," were all at academic research centers: the University of California, Los Angeles; the Stanford Research Institute; the University of California, Santa Barbara; and the University of Utah.

The ARPANET was a blue-sky research project that, ostensibly, would allow for greater (and more resilient) communications among decision-makers during a nuclear strike. While they sold their project to the generals this way, the academics behind the ARPANET then spent the next twenty years evolving the network into a system that better suited their own needs: a distributed, nonhierarchical computer and communications network that facilitated discussion and exchange among the research and scientific community. The ARPANET evolved into the Internet we recognize today not as a populist or mass-market communications system, but as an electronic playground where a priesthood of academics could play and exchange ideas.

This elitist focus showed in the Internet's maturity. All of the various Internet protocols that developed, from the most obvious, such as FTP (File Transfer Protocol) and TCP/IP (the basic building block of the Internet itself), to the most recent and seemingly sophisticated, such as newsgroups, Gopher (the first true Internet search) and even email—all were complex to set up, unfriendly to nonexpert users, and, frankly, boring and utilitarian. Even as computers became personal, even as computing itself was becoming colorful and democratized, the Internet remained stubbornly aloof, sequestered in the ivory towers of academia.

The Internet, in short, needed its own GUI revolution, that application/user interface innovation that would make the Internet user-friendly just as the graphical user interface had done with computing itself. The World Wide Web arrived just at the right time, and provided this exact paradigm shift just when it was needed.

The web came in 1990, just as Windows was beginning to take computers into the majority of the world's homes and offices, and just as the computer mouse and the graphical icon were making computing point-and-click intuitive for everyday people. The web lived in this world. You navigated the web with a mouse, you clicked on links, and the whole thing moved with the innate, logical simplicity of how human thought seems to work: jumping from one idea or association to another, flowing backward and forward in the direction of idea and inspiration, reference and retort. The web took the fundamental concept of the Internet (connecting computers together) and made it manifest through the genius of the hyperlink. One website linked to another. One idea linked to another.

This metaphor of the link made the whole conceptual idea of the Internet, of linking computers together, of linking people's minds together, of linking human thought together, finally and wonderfully real.

And yet, the web itself was *still* a child of academia. It remained a researcher's dream of a scholarly utopia. It is well known that Tim Berners-Lee invented the web while he was employed at CERN, the great multinational scientific research institution in Switzerland. As the Internet was born in the midst of a great scientific effort to win the Cold War, the web was born in the midst of a great scientific effort to reveal the secrets of the Big Bang.

Berners-Lee saw his new Internet protocol as an improvement on top of the existing structure of the Internet itself. He built the web upon previous conceptual and philosophical notions (hypertext, cyberspace, collaboration) to create what was really a new medium. In his Usenet post announcing the web, Berners-Lee declared, "The WWW project merges the techniques of information retrieval and hypertext to make an easy but powerful global information system."[3] But he still envisioned it—at heart—as a research medium, a way for the hundreds of CERN scientists from all around the world to share their data, disseminate their ideas, and collaborate on research.

Again, from his announcement post:

> The WWW project was started to allow high energy physicists to share data, news, and documentation. We are very interested in spreading the web to other areas, and having gateway servers for other data. Collaborators welcome!

The collaborators Berners-Lee was calling for were imagined as fellow researchers and academics. The web, for all the structural ways that it would eventually prove friendly to the average computer user, was still intended for the priesthood, not the masses.

There was one, final, catalyzing event that had to happen before the web—and with it, the Internet on the whole—could go mainstream. There was one more innovation necessary before the average user would be invited to join the computer revolution en masse, and we could create a world with Amazon and smart televisions and app stores and self-driving cars and cat memes.

That one more thing would, in fact, come from a research institution, but it would serve to wrest the Internet and computers themselves from the privileged clutches of academia forever and thrust them into the loving embrace (and eventually the pockets) of average users like you and me.

1

THE BIG BANG

The Mosaic Web Browser and Netscape

Netscape Communications Corporation held an initial public offering, or IPO, on August 9, 1995. Netscape shares were originally to be priced at $14 per share, but at the last minute the price was lifted to $28 per share. When the markets opened at 9:30 A.M. Eastern Time, Netscape's stock did not open with it. Buyer demand was so great that an orderly market could not immediately be made. Interest from individual investors was so overwhelming that callers to the retail investment firm Charles Schwab were greeted by a recording that said: "Welcome to Charles Schwab. If you're interested in the Netscape IPO, press one." At Morgan Stanley, one retail investor offered to mortgage her home and put the proceeds into Netscape stock. The first Netscape trade did not hit the ticker until around 11 A.M. The price of that first trade was $71, almost triple the offer price.

Over the course of the day Netscape, with the ticker symbol NSCP, reached $75 before ending the day at a respectable $58.25. Netscape had only existed as a corporation for sixteen months. Since its inception, it had generated revenues of only $17 million. It had nothing in the way of profits on its balance sheet. But at the end of trading that day, the stock market valued the company at $2.1 billion.

These days we're used to embryonic technology companies

debuting on the stock market to soaring valuations, but in August of 1995, such an event was almost unheard of. The financial press was in awe, if skeptical. On its front page the next day, the *Wall Street Journal* said, "It took General Dynamics Corp. 43 years to become a corporation worth $2.7 billion. . . . It took Netscape Communications Corp. about a minute."[1] Plenty of commentators were shocked that a company that had yet to make any sustained profit could be valued so highly. Still others were puzzling over what this "Internet" thing even was, and why it was making people rich. As August 9 also happened to be the day that Jerry Garcia of the Grateful Dead died, a joke made the rounds on Wall Street: What were Jerry Garcia's last words? Answer: "Netscape opened at what?"

A lot of the chatter was about the sudden, unprecedented and remarkable creation of wealth. Cofounder Jim Clark's 20% stake in the company was worth $663 million on the day of the IPO. Early Netscape employees were worth millions of dollars (on paper at least), including the company's baby-faced, twenty-four-year-old cofounder, only a few months out of college, who was suddenly worth $58 million.

A few short months later, in December 1995, Netscape's stock price would hit $171 a share, more than six times the price at the IPO. A few weeks after this milestone, that same twenty-four-year-old, Marc Andreessen, would grace the cover of *Time* magazine.

There are occasionally events that signal the arrival of a new force in culture (say, the Beatles on *The Ed Sullivan Show*) or serve as the demarcation line between historical eras (September 11, 2001, for example). The Netscape IPO was just such a moment in time. Today, young twenty-somethings dream of coding their way to billion-dollar fortunes. Today, the phone in your pocket is more powerful than every computer involved in the moon landing. Today, it's possible to know, in real time, what your high school crush had for lunch. Netscape set the groundwork for this reality. The Netscape IPO was the big bang that started the Internet Era. That picture of a barefoot Marc Andreessen on the cover of *Time* was what started young geeks dreaming of Silicon Valley riches. Netscape would not define the Internet Era—or even survive it—but it was the first of its kind, and in many ways it was the template for all the people and companies that would follow.

■

THE MODERN WEB ERA began in Champaign, Illinois. The University of Illinois at Urbana-Champaign is world-famous as a leading research institution in the field of computing. The ORDVAC and ILLIAC, two of the earliest computers in the world, were built there in 1951; the university was granted Unix license number one by Bell Laboratories in 1975; and in 1985, the National Center for Supercomputing Applications (NCSA) was established there. In the famous science fiction movie *2001: A Space Odyssey,* the homicidal HAL 9000 computer states that he "became operational" in Urbana, Illinois, on January 12, 1992, partially as a nod to the university's prominence in the field.

When the National Science Foundation took over the operations of the Internet in the 1980s, the University of Illinois was a key part of the Internet "backbone," that superstructure of digital pipes that allowed the network to function.[2] By 1992, when the superfast T3 network was launched as the successor backbone for the Internet, the NCSA and the university were sitting on some of the fastest computer connections in the world. In other words, by the early 1990s, there wasn't a better place in the world if you wanted to be swept up in the revolution of the World Wide Web.

It helped that the NCSA was relatively flush with cash and resources in the early 1990s. It had gotten a large amount of funding thanks to the recently passed High Performance Computing Act of 1991, more commonly referred to as the "Gore Bill."* All the wired infrastructure, all the superfast computing machines and the small army of undergraduate and graduate students the NCSA employed to assist with research projects, were paid for and paid by, in part, the government.

---

* So named because Senator Al Gore introduced and championed the legislation. One could insert a joke here about Al Gore inventing the Internet, but the Gore Bill played a crucial role in the early experiments we're about to discuss, as Marc Andreessen himself would later credit.

"NCSA was heaven," remembers one of the students working there in the early nineties, Aleks Totic. "They had all the toys, from Thinking Machines to Crays to Macs to beautiful networks. It was awesome."[3]

Another student programmer, Jon Mittelhauser, would remember, "We were all just kids hanging out in the basement of what was called the software development group."[4] The professors who ran the research programs that were the NCSA's bread and butter assigned the projects, and the pool of "kids" in the basement coded away to the profs' specifications.

In 1992, one of those kids was a twenty-one-year-old by the name of Marc Andreessen. Born in Cedar Falls, Iowa, on July 9, 1971, Andreessen grew up in New Lisbon, Wisconsin (pop. 1,450),[5] where his father was a feed salesman and his mother was a shipping clerk at Lands' End. Computers fascinated Andreessen when he was a child, and he taught himself how to program at an early age. But he was no prodigy. Built large, at six feet two, with a loud, excitable personality, he was not exactly a wallflower, and it set him apart. Another NCSA student programmer named Rob McCool remembers of Andreessen, "All of the [computer science students] I'd come across were all quiet, kind of nerdy types. And here's this gigantic Scandinavian guy with a purple computer and he's wild-eyed and telling me about all this stuff that's gonna be great."[6]

Andreessen was voluble and enthusiastic, but he also had an antiauthoritarian, independent streak that his peers came to appreciate. When a research team Andreessen and McCool were part of got hung up on a coding problem relating to an assigned project, Andreessen simply junked the existing framework and hacked together his own solution. "And I was like, 'Dude, really? Can you do that?'" McCool remembers. "And he was like, 'Yeah, well, my boss hasn't noticed yet.'"[7]

Andreessen had joined the NCSA as a part-time student programmer, doing menial coding work for $6.85 an hour. The researcher who hired Andreessen was Ping Fu, who had had a hand in the groundbreaking "morphing" computer graphics featured in the recent feature film *Terminator 2: Judgment Day*. Andreessen's main task at the NCSA was coding Fu's visualization projects. But what really caught Andreessen's imagination during those hours in the NCSA basement, the computing technology that he was telling McCool and others was "gonna be great," was that latest and greatest thing on the Internet: the World Wide Web.

With the NCSA's fast computers and even faster Internet connections, Andreessen and the other kids in the basement were perfectly positioned to catch the wave of the Web when it took off. In fact, the NCSA was just the sort of academic research organization that Tim Berners-Lee was fervently hoping would adopt his invention. At this point in the web's development, Berners-Lee had just recently open-sourced his project to the world, in the hopes that he could "let a thousand flowers bloom" by inviting others to contribute to the project's development. At the time, there were maybe a couple hundred software developers in the entire world experimenting with the web, and they all hung out and exchanged ideas with Berners-Lee on a Usenet newsgroup called WWW-Talk.

In November 1992, there were only a few dozen WWW servers in the world. By the end of that same month, one of them happened to be at the NCSA, courtesy of Marc Andreessen.[8] On November 16, 1992, Andreessen showed up in the WWW-Talk message group for the first time, joining the various conversations about HTML, web servers and web design and generally volunteering to pitch in on the grand project of moving the web forward.[9]

Moving forward meant a better web browser. A browser is a software application that allows a user to both navigate and view the web. Berners-Lee himself had coded the first browser back when he had invented the web. But, as a part of his new crowdsourcing efforts, he had thrown the door open to anyone who wanted to try their hand at coding a better one. Dozens of developers around the world accepted the invitation, and several of them turned out to be students around the same age as Andreessen. At the University of Kansas, several students created the text-based Lynx browser. Pei-Yuan Wei developed the ViolaWWW browser while pursuing a degree at UC Berkeley. If you wanted to make a splash in the early web community, the way to do it was to code and release a better browser, and Marc Andreessen wanted to make a splash.

Andreessen himself would later describe the early web this way:

> PC Windows had penetrated all the desktops, the Mac was a huge success, and point-and-click interfaces had become part of everyday life. But to use the Net you still had to understand Unix. . . . And the current users had little interest in making it easier. In fact, there was a definite element of not wanting to make it easier, of actually wanting to keep the riffraff out.[10]

Andreessen's big idea in the winter of 1992–93 was to let the riffraff in. He wanted to release a simpler, more user-friendly browser. He wanted it to be point-and-click and windowed. He wanted to make the web look familiar to someone who was comfortable using a personal computer, as opposed to the Unix workstations most of the researchers on the web were used to. And, crucially, he wanted the web to look as sexy as it felt to people like him who were enthusiastic converts. He wanted to add pictures. Says Aleks Totic: "[Andreessen] was like, 'Oh, there could be newspapers on the Net and all this information can be out there for everyone. How phenomenal could that be?' "[11] In short, Andreessen had a vision for the web in which someday everything would be possible: graphics, news, commerce, even cat videos.

So, Andreessen turned his special brand of infectious enthusiasm on his fellow NCSA coders. The first person he targeted was his colleague Eric Bina. Bina was older than Andreessen (almost thirty) and a full-time, salaried NCSA employee. Bina was also a much better programmer than Andreessen was. Bina initially begged off the project, but Andreessen's enthusiasm and persistence eventually won him over. The "browser project" that Andreessen and Bina undertook began sometime in December 1992. Bina wrote the majority of the original code, but the features were also what made their browser such a leap forward, and it was Andreessen who was coming up with the features.

In a little over a month of nearly round-the-clock coding, they had their browser ready. It was called X Mosaic. On Saturday, January 23, 1993, the official "0.5" version of the browser was posted to the Internet on the NCSA's servers. The accompanying release note from Andreessen himself said:

> By the power vested in me by nobody in particular, alpha/beta version 0.5 of . . . X Mosaic is hereby released.

The last line of the message was the FTP address telling others where they could go to download and install the browser themselves. Within days, no less a web authority than Tim Berners-Lee forwarded and endorsed Andreessen's announcement:

> An exciting new World-Wide Web browser has come out, written by Marc Andreessen of NCSA.

This browser was called "X Mosaic" because it was designed to work with X Window, a graphical user interface popular with users of Unix machines. It was designed for the computers that researchers and academics used. In other words, it was preaching to the already-converted web choir. And that was not what Andreessen was after, of course. Using X Mosaic as a proof of concept, he turned his enthusiasm on others in the NCSA basement to get them to write versions of his browser for the computers that the riffraff used.

NCSA's young programmers signed on to program these versions, each according to his own platform of choice. Jon Mittelhauser and Chris Wilson developed the PC version. Aleks "Mac Daddy" Totic and Mike McCool wrote the Macintosh port. And since X Mosaic handled only the consumption end of the web experience, the growing team thought it would be a good idea to tackle the delivery end as well. Thus, McCool's twin brother Rob wrote Mosaic web server and publishing software that would eventually be released alongside the browsers.

The kids in the basement did their thing and then released it out into the world. That was how the web worked in 1993; that was what

Tim Berners-Lee had hoped would happen. If someone had a better way of doing things, they coded it up and made it available for other people to try. If others liked it, they downloaded it. If they didn't, well, they didn't. And if these users had problems, found bugs, had ideas for improvements or wanted to contribute new features, then they got in touch with the creators over email or the Usenet message boards and bitched about it. The kids at the NCSA, surrounded by empty pizza boxes and soda cans, released updated versions—and then maybe a week later they released another updated version based on user feedback.[12] The process was very communal and very real-time.

■

WITHIN EIGHTEEN MONTHS, Mosaic was the biggest thing on the web, and probably the biggest thing on the Internet at large. In January of 1993, shortly after Mosaic launched, the number of websites in existence was in the hundreds. By the end of 1994, the number of websites in the world had passed tens of thousands.[13] In a similar time frame, the number of web hosts had risen tenfold.[14] In a way, one could argue that Mosaic helped make the web, and vice versa. As the first browser designed for the common computer user, Mosaic had a symbiotic sort of chicken-and-egg relationship with the web. For millions of PC and Mac users, Mosaic was their first glimpse of the web. Once they saw what the web could do, they wanted to go off and code their own websites.

Within those first eighteen months of launch, Mosaic probably delivered 3 million browsers into users' hands.[15] That may seem like a small number, but then, there probably weren't many more than 3 million people on the web before that point. Toward the tail end of 1994, Mosaic was adding as many as 600,000 new users every month. It is safe to say that by that point the vast majority of people surfing the web did so via a Mosaic web browser.

The key innovation of the Mosaic browser was Andreessen's insight that in order to make the web sexier, he simply had to release a browser that enabled the sexiness he imagined. On February 25, 1993, mere weeks after Mosaic's initial beta launch, Andreessen was on the WWW-Talk message boards making a proposed addition to HTML of an "inline" image tag that would allow for images to be coded directly into web pages. Prior browsers opened images—and really any non-HTML file type—as a separate window. Inline images would make web page design more akin to the page layout of a magazine or newspaper.

Adding color and sexiness to the web was part of what made Mosaic take off, and part of what made the web take off at exactly the same time. But even the web's creator was among those who felt that Andreessen's penchant for multimedia was a little much. Andreessen later admitted, "Tim [Berners-Lee] bawled me out in the summer of '93 for adding images to the thing."[16]

■

"HE ONLY WANTED TEXT," Andreessen has said of Berners-Lee's objections when they finally met face-to-face at the World Wide Web Wizards Workshop in Cambridge, Massachusetts, the first true developer conference. "He specifically didn't want magazines. He had a very pure vision. He basically wanted it [the web] used for scientific papers. His view was that images are the first step on the road to hell. And the road to hell is multimedia content and magazines, garishness and games and consumer stuff. I'm a Midwestern tinkerer type. If people want images, they get images. Bring it on."[17]

For his part, Berners-Lee has denied that images discomfited him. "Of course we did approve of images, in fact we had images on the Web before anybody else," he has said. But then he adds, "Like diagrams in talks for example."[18]

Years later, even Mosaic cocreator Eric Bina would admit that he had reservations about adding images and multimedia to the web. At the time, he was mainly concerned about bandwidth issues (this was the era of dial-up modems; images could take entire minutes to load onscreen) but he was also worried that he and Andreessen were opening the floodgates to frivolity and junk. "And I was right! People abused it horribly," Bina said later. "But Marc was also right. As a result of the glitz and glitter, thousands of people wasted time to put in pretty pictures and valuable information on the Web, and millions of people use it."[19]

■

THE MILLIONS OF DOWNLOADS to users around the world meant that Mosaic was probably the most successful software product ever designed for—or released on—the Internet up to that point. By the end of 1994, it was clear that the World Wide Web was rapidly taking over the Internet at large. For the millions of Mosaic users, the web almost *was* the Internet. But then, those millions of users were not exclusively the academics and researchers the web had been designed for. Increasingly, they were also home computer users; business computer users; the uninitiated; the uninvited; the riffraff. Mosaic had become the most successful project in computer science by leaving the computer scientists behind and appealing to the mainstream. *Fortune* magazine named the Mosaic browser one of its products of the year (alongside the Wonderbra and Mighty Morphin Power Rangers), writing, "This software is transforming the Internet into a workable web . . . instead of an intimidating domain of nerds."[20]

As these things tend to go, the popularity of Mosaic, and the nerd celebrity of the Mosaic team especially, started to create friction within the NCSA. The higher-ups at the center had originally thought of Mosaic as just another software project, very much fitting into their larger purview as a computing research institution. But, over the course of 1993, the status of the Mosaic browser project changed; it became a major NCSA priority. At first the NCSA bigwigs didn't seem to "have any clue who we were, and we liked it that way," said Jon Mittelhauser, but once Mosaic took off, "We suddenly found ourselves in meetings with forty people planning our next features, as opposed to the five of us making plans at 2 AM over pizzas and Cokes. Aleks, who had basically done the Mac version [on his own], suddenly found out that there were three or four other people working on it with him, according to the NCSA. And they were like his bosses, telling him what to do."[21]

Chris Wilson was another of the NCSA student programmers who would later go on to work at Microsoft and develop the first Internet Explorer browsers. "I think that Marc and some of the other guys there really wanted to see NCSA just drop everything else like a hot rock and go totally support the web and scale up to do it," Wilson says. "If you were in a startup and you saw one of your products getting so much attention and having so much potential, absolutely you'd figure out how to do that, right? You'd go mortgage your house."[22]

In fact, the Mosaic team already was functioning like a software startup in all but name, while the NCSA was still thinking of the browser as a glorified research project. Increasingly, this conflict in vision extended to the very structure of the core team of programmers. The part-time student coders were muscled out as the higher-ups assigned seasoned, full-time employees to the project. It was suggested to Andreessen especially that, for the good of the project, he should step aside and let more experienced hands take over. "Don't you think it's time to give someone else a chance to share the glory?"[23] he was asked.

In December of 1993, Mosaic and the web made the front page of the *New York Times*. NCSA director Larry Smarr was pictured and quoted: "Mosaic is the first window into cyberspace," he said.[24] Neither Marc Andreessen nor anyone else on the Mosaic team was even mentioned.

"[Andreessen] had to lead at NCSA," says Aleks Totic. "And if he couldn't lead, he had to leave."[25]

Andreessen was due to graduate that same December. He didn't even bother to pick up his diploma. By the end of 1993, just a year after launching the Mosaic browser, Marc Andreessen was in Silicon Valley looking for work.

■

THE SILICON VALLEY that Marc Andreessen found himself in by early 1994 was actually at a historical low ebb, considering what was in store. The short but sharp recession of 1990–91 hit the technology industry hard. PC shipments fell by 8% in 1991, the first such drop in recorded industry history.[26]

"I thought I had missed the whole thing," Andreessen would later say of his arrival in California. "The overwhelming mood in the Valley when I arrived was that it was done. The PC was done, and by the way, the Valley was probably done because there was nothing else to do."[27]

Forget the Valley, in 1994, what was the something else that Marc Andreessen could do? To us now, the answer is obvious: form a startup; get venture capital backing; release a product; gain millions of users; go public; become a billionaire. This is only the obvious path to modern minds because of the "something else" that Marc Andreessen would do in 1994: cofound Netscape, the first true Internet company, the first real "dot-com." At the time, there was no template for Marc Andreessen to do a web startup, because Marc Andreessen hadn't created that template yet.

"I had some idea that I wanted to be part of a new company," Andreessen says, "but I didn't even know what a VC [venture capitalist] was."[28]

■

JIM CLARK IS FAMOUS in Silicon Valley history for having founded three different billion-dollar companies. By the beginning of 1994, Clark was just departing billion-dollar company number one: Silicon Graphics (SGI). Jim Clark's tenure at Silicon Graphics was not ending on a happy note. Despite being the founder, despite being largely responsible for the development of modern computer-aided design and computer graphics (those dinosaurs in *Jurassic Park*? You can thank Silicon Graphics for those), despite turning SGI into a multibillion-dollar publicly traded enterprise, Clark found himself edged out of his own company.

And that wasn't the worst of it. What really stuck in Clark's craw was the fact that he wasn't filthy rich. Clark believed he had built SGI into a technology powerhouse that rivaled the likes of Microsoft and Oracle. And yet, he had nowhere near the wealth of Bill Gates or Larry Ellison to show for it. The need to raise venture capital in the early years of SGI's development had repeatedly diluted Clark's ownership share so that, despite Silicon Graphics' billion-dollar valuation, Clark had a net worth of only about $20 million. He had billionaire envy.

Clark told SGI and the press that he wanted to start a new company. He resolved that this time he would do things his way, and he would hold on to enough equity to become a billionaire. The trouble was, Clark didn't know what his new company would do, exactly. He had some vague ideas about creating software or hardware for interactive television, what was being called the information superhighway. The information superhighway was supposedly the next big thing, and that was exactly what Clark wanted to be a part of. He even went so far as to have exploratory meetings with companies like Time Warner and Nintendo. After all, if interactive TV was the next big thing, then you could do worse than have the founder of Silicon Graphics helping you build the set-top boxes.

But really, Clark was just casting around for anything that would give him a second act. And this meant that he was open to ideas. Any ideas. He turned to his friend Bill Foss, a veteran Silicon Valley engineer; did Foss know anyone smart Clark could talk to?

"Well, what about Marc Andreessen?" Foss asked Clark. "He just moved to Palo Alto from Illinois."[29]

By way of explaining who Andreessen was, Foss loaded a version of the Mosaic browser onto Clark's computer. Clark must have been impressed; shortly after his first session using Mosaic was over, he sent the following note to Andreessen's personal email address:

> Marc:
>
> You may not know me, but I'm the founder and former chairman of Silicon Graphics. As you may have read in the press lately, I'm leaving SGI. I plan to form a new company. I would like to discuss the possibility of your joining me.
>
> Jim Clark.

Sometime in early 1994, Jim Clark and Marc Andreessen met at 7 A.M. at a coffee shop in Palo Alto called Caffe Verona. Andreessen had found a job at a Palo Alto–based company named Enterprise Integration Technologies, working on Internet security products. Even while gainfully employed, Andreessen certainly knew who Jim Clark was, and he was very interested in whatever new venture he might be cooking up.

Andreessen would later remember that it was the first time he had been up that early in several years. Clark told him that he was looking to start a new company. He didn't know what kind of company it would be yet, but he was looking for people to help him figure it out.[30] Clark must have been impressed with Andreessen, because he invited the young engineer to join a small group of Clark's trusted associates, including Bill Foss, who would meet on a regular basis at Clark's house to kick around ideas.

During one such confab, at about one in the morning in late March 1994, Clark said to Andreessen simply, "You come up with something to do and I'll invest in it."

"Well, we could always build a Mosaic killer," Andreessen told him.

■

MORE THAN ANYONE ELSE in the world, Marc Andreessen knew that the next big thing was the World Wide Web. The information superhighway might have been what all the smart, big money people like Clark thought was going to be next, but Andreessen understood that Clark didn't have to chase dreams of interactive TV or cut deals with cable companies—the future was already here, and millions of people were already using it.

It came down to simple numbers. Andreessen showed him that users of the web were doubling every few of months at that point—absolutely exponential growth. Clark didn't know how someone could make money off that growth exactly, but he figured with numbers like that, there had to be a way. Andreessen had proven with Mosaic that a web browser was a pretty darned good way to piggyback on that growth explosion. As Clark came around to this point of view, the notion that thrilled him the most was the idea that they could pounce on this opportunity first. Let the rest of the world develop the information superhighway. He and Andreessen would deliver it before anyone else was any wiser.

What would eventually become Netscape was formally incorporated as Mosaic Communications Corporation on April 4, 1994. The first order of business was locking down a software team capable of coding a better browser. Andreessen had been careful to keep in touch with his former colleagues back in Illinois, so it was just a matter of getting the band back together.

"Marc basically sends mail, says, 'Hey, I met Jim Clark. He's a cool guy. He's looking to start up a company. And I'm talking with him about what we should do,'" says Jon Mittelhauser.[31]

"One thing led to another," remembers Aleks Totic. "And he said, 'We're not going to do a Nintendo network, I think we're going to do the web.'"[32]

Clark and Andreessen flew back to Urbana-Champaign and checked into the University Inn. They met their quarry (the original Mosaic team: Eric Bina, Aleks Totic, Jon Mittelhauser, Rob McCool, plus two additional outside engineers, Chris Houck and Lou Montulli) at a pizza place near the University of Illinois campus. Clark offered the men identical $65,000-a-year salaries, one week's paid vacation in Tahiti on Clark's own yacht, and, more important, 100,000 shares of stock in the new company.

"So, we go out to this place and they're basically like, 'Yeah, let's do Mosaic, except let's make a company out of it,'" says Rob McCool. "He [Clark] has this Jedi Mind Trick speech where he brings us all upstairs and we all come down saying, 'Yes, we're going to make a company! It will be great!'"[33]

Clark told the team: "Within five years, if things go the way I hope they will, it is my objective that you make over ten million [dollars]."[34]

Clark typed up identical agreement letters on his laptop and had them printed on the University Inn's fax machine. The whole team signed on and retired to a bar named Gully's to celebrate.

"We didn't really know much about Jim Clark," says Aleks Totic. "But we trusted Marc. He gave us all these papers to sign. We just met him for one night. Next morning, we all walked in and quit.

That was on Thursday. On Saturday, we were in California picking out apartments."[35]

Today, recent college graduates from around the world dream of heading to Silicon Valley and finding their fortune. The original Mosaic crew was the first to make this journey. They didn't know they were the vanguard of a newfangled gold rush. They were, literally, corn-fed midwesterners. They were used to making six bucks an hour for their coding and had little inkling that software development could pay much more than that. When Jim Clark dangled a high-five-figure salary in front of them, they almost thought he was joking.

But the midwestern kids showed up in California to find that 11,699 square feet of real office space had been secured for them above a Mexican restaurant at 650 Castro Street, the main drag in the town of Mountain View. Work quickly commenced on a new web browser that would be better than Mosaic. Mac, Windows and Unix versions of the new browser would be developed simultaneously. The browser code and the server code would be rewritten, with a focus on greater speed, greater stability and better features. In other words, this was to be a proper product, not just a research project.

Their first effort had been a bit of a hobby, a lark project. At the NCSA, "we were students; we were just having fun," Mittelhauser recalls. "We had no thoughts about quality, really. That was the coolest thing about doing Netscape after Mosaic. We literally started from scratch and were able to avoid many of the same mistakes (while of course making new ones)."[36] This time they would do it better and get it right. All hands were tasked with speedily producing what the team of young coders had dubbed "Mozilla," suggesting that the new browser was a monster set to devour their previous brainchild, Mosaic.

■

THE COMPANY THAT WOULD BECOME Netscape was the first web company, the first true dot-com. In so many ways, it blazed a trail and set a template for what we think of as a modern technology startup. Details we take for granted about the modern tech industry can trace their roots to the Netscape story, whether accidentally or by design. One of the ways this manifested was in the corporate culture of the young company. Everything was about speed, about what Jim Clark would call "Netscape Time," but would later be widely adopted by the media as "Internet Time." For most of the twentieth century, the "product cycle" was something that happened in a comparatively leisurely, plodding, measurable pace. But as has become common to the point of cliché over the last twenty years, in the Internet Era, change—whether to products, industries or entire economies—would come literally overnight.

With Mosaic, the NCSA kids had stumbled upon something that truly represented a new method of software development, a new ethos for product development. Software, at the time, meant floppy discs or CDs sold in cardboard boxes at retail. Jim Clark came from the world of machines and hardware, where development schedules were measured in years—even decades—and where "doing a startup" meant factories, manufacturing, inventory, shipping schedules and the like. But the Mosaic team had stumbled upon something simpler. They had discovered that you could dream up a product, code it, release it to the ether and change the world overnight. Thanks to the Internet, users could download your product, give you feedback on it, and you could release an update, all in the same day. In the web world, development schedules could be measured in weeks.

It was this new paradigm for product development, more than anything else, that was Netscape's first contribution to the modern idea of "a startup." Marc Andreessen described it this way: "You keep kicking versions out the door, making them better. Any individual product is less important than the basic idea. If a beta turns people off, you put out a beta that turns them back on."[37] Jim Clark eagerly embraced this new way of doing things. "You didn't build some physical thing, move it down an assembly line, box and shrinkwrap it, and stick it on a store shelf," Clark wrote in his autobiography. "You conceived of it in your head, produced it in a computer, and tossed it up for grabs on the Net."[38]

This new paradigm demanded an almost 24/7 work schedule,

another now ubiquitous feature of Silicon Valley that Netscape would enshrine. During the period of the new browser's development, a young programmer named Jamie Zawinski regularly posted to an online diary. These entries (which would be considered blog posts today) captured what it was like to be a member of the team. He described working for as many as thirty-nine straight hours, catching catnaps under the desk in his cubicle, missing meetings because of fatigue, hoping to catch his "second or third or eighteenth wind."[39]

Software engineering has always been a pursuit that lends itself to intense bouts of work, long bursts of productivity when you come up for air and realize you've been coding for days straight. In a way, we can't blame Netscape for the high-intensity template that it would bequeath to our collective understanding of startups. Even though so many of the breathless news clippings from the time focused on the all-nighters and the frat-house hijinks of the Netscape offices, these were, after all, young men fresh out of college. That's just what they knew.

"We [were] working around the clock because that's what you used to do before," says Aleks Totic. "Four years later, five years later, the entire valley [would] be living the same lifestyle. But those people actually have lives. We really didn't have any lives outside of the office so of course we're going to be at the office all the time! I mean, I had no furniture. Why should I ever go home?"[40]

Lou Montulli says, "The press just take what they think is most interesting, juicy and fascinating out of their limited time and they publicize that. Especially post-Netscape, in 1998, 1999, every startup was trying to do the things they read about in magazines." Montulli admits that his own schedule was inhuman at the time. "I would catch about four or five hours of sleep at the office . . . wake up and do another 20 hours and then go home and sleep for about 12 or 15 hours and then start the whole cycle again. I wouldn't recommend doing that to your average startup. Unfortunately, a lot of startup people think that that's the way it should be done because of all the publicity we had."[41]

Other features that now define Silicon Valley startups include the informal working environment and the insane perks that companies seem to dole out freely. Netscape pioneered this informal work culture as well, but in retrospect you have to wonder if it was all just a matter of motivating twenty-something male software engineers.

Netscape had foosball and air hockey and networked computer games and anything else postcollege bros thought was cool circa 1994. The most notorious intracubicle competitions were the bouts of chair football, gladiatorial contests pitting contestants against each other while riding atop their rolling desk chairs. Chair football was brutal, and sometimes even bloody. "We probably took out about ten chairs [because of] that game," recalled Bill Foss, who had joined the company as an advisor.[42]

"There was a huge movement to play multiplayer Doom [in the office]," remembers Rob McCool, referencing the then-popular first-person shooter video game. "It got to the point where they started having to threaten disciplinary action; making policies of no gaming before 5 P.M. and that kind of stuff."[43]

One person who was seldom participating in any of these hijinks was Marc Andreessen. Nor was he participating in the all-night coding sessions. Now, in California, at a real company, developing a real product, Andreessen's role was different. Jim Clark had made good on his offer to build a company around Andreessen's ideas. From the earliest days, Andreessen was referred to as the new company's cofounder.

Andreessen had been dragooned into becoming the public face of the new enterprise. Rosanne Siino, a PR manager who had followed Clark over from SGI, knew she had a good story on her hands. "I thought, I've got the Internet, which is hot; I know I can make a big deal out of that. I've got Jim Clark, who is hot," Siino remembers. "And then I've got this twenty-two-year-old wonder kid. No matter what, it's going to get a lot of coverage."[44] Soon, Andreessen and Clark were being featured like a dynamic duo in articles like *Fortune* magazine's "25 Cool Companies" list. *Fortune* dubbed Andreessen "the hayseed with the know-how."[45] The *San Jose Mercury News* featured Andreessen in an article titled "He's Young, He's Hot, and He's Here." Toward the end of 1994, *People* named Andreessen one of its "Most Intriguing People," alongside a young golfer named Tiger Woods.[46]

At the same time, Andreessen and Clark were settling upon the business strategy the new company would pursue. To this end, the pair increasingly looked to the obvious inspiration at that time: Microsoft. Microsoft's operating systems had a monopoly hold on the personal computer market. DOS and Windows were the platforms that the vast majority of the computer world had to build off of and exist on. If you were a programmer and you wanted to create a program that would reach the greatest number of users, then you worked with Bill Gates's platform. Sometimes that meant paying a toll to Gates, and sometimes it didn't. But either way, you played ball with Microsoft or you found yourself and your program relegated to the hinterlands of the computing world.

Andreessen and Clark began to think of the web browser as a sort of platform for the web. Why couldn't the web browser be the DOS/Windows for the Internet? The key was to become the market standard, which meant being first. But becoming a platform also meant enticing developers to develop for your platform. Almost from the very beginning, Andreessen and Clark wanted their web browser to enable an ecosystem that other programs, and even other companies, could be built off of. Throughout its life, Netscape would embrace open-source culture and practices. If they were the first browser to introduce support for an innovation, they didn't make that advance proprietary. They allowed the new feature to be used by others, hoping it would become standard, and hoping they would get credit for the innovation and for being first. A good example of this was the Secure Sockets Layer (SSL) technology, which Netscape would pioneer. This is the encrypting technology that makes secure interactions on the web possible. Netscape's browser would be the first to feature this technology, but Netscape left the underlying standard free for others to use and support. This open attitude toward the technology is what allowed the first ecommerce activities to begin flowering across the web. Netscape benefited as the underlying platform that was the most trusted and valued by users. Netscape also eagerly supported and incorporated the advances of others—for instance, the Java programming language when it emerged. Netscape would even encourage others to build add-ons and plugins that would interact with Netscape's own software, adding features and functions that Netscape itself couldn't dream up.

Throughout the Internet Era, company after company would become obsessed with the idea of creating or owning a platform. If you are a platform, you can create an ecosystem of developers and software and apps all dependent on the underlying platform. To own a platform is to own the ball field, the rule book, the turnstiles, and the broadcast rights. Netscape did not originate the obsession with platforms, but it would provide the template.

While he and Andreessen were busy hashing out product and strategy, behind the scenes Clark was busy forming a company that would be ready for the big leagues. Experienced engineering managers were brought in to oversee the development team. Clark knew he wanted a world-class CEO (he himself was content to be chairman of the board), and he swung for the fences, setting his sights on Jim Barksdale, the much-in-demand former vice president and chief operating officer at FedEx, and currently the CEO of McCaw Cellular.

Clark also raised capital for the company, though he put that off as long as he could. Having been burned by his SGI experience, for many months Clark funded operations out of his own pocket, keeping a firm grasp on his sizable equity stake. When the time finally came for investment, Clark's reputation secured funding on very favorable terms from the premier venture capitalist in Silicon Valley, John Doerr, a partner at Kleiner Perkins Caufield & Byers. Kleiner Perkins (as the firm is generally known) had directed venture capital funding for such early technology giants as Compaq, Intuit and Sun Microsystems, and Doerr himself would go on to fund web companies such as Amazon and Google, among many others, in the coming years.

The people were in place. The funding was in place. The browser was deep into development. The final question was an important one: how to make money? Andreessen and Clark eventually settled on a seemingly radical strategy: the product would be free. Well, in a winking, knowing sort of way. Upon release, the web browser would be available for anyone to download so-called beta versions ("beta" means an early version of the software; a work in progress). However, if you wanted to own the standard version of the software—the final one, with all the bells and whistles and customer support—it would cost $39. (Even this was fungible. Anyone would be able to download the full version of the software on a trial basis for ninety days. After that, you were supposed to pay up.)

"At the time, it was a crazy idea, to build this software but just give it away," says Rob McCool. "They were going to give away the browser and charge a lot of money for the server."[47]

"Essentially, the razor and razor blades model," says Netscape engineer Lou Montulli.[48]

This was a savvy move. At the time, everything on the Internet was free. If he wanted to be among the first to ask users to pay for web-based software, Andreessen knew he had to tread lightly. The idea was to hook users on the free beta version, and then to ask them to pay up for the finalized product, a "pro" version. If corporations wanted to get into the act, they would have to pay up—to the tune of thousands of dollars—for the servers to make the web work within their organizations. Being free would help the browser gain market share, which was the sine qua non of his platform strategy. If the new browser could quickly match Mosaic's then 90% market share, then they would become the de facto standard that all other browsers would be measured against.

"It's basically the Microsoft lesson, right?" Andreessen asked. "If you get ubiquity, you have a lot of options, a lot of ways to benefit from that. You can get paid by the product that you are ubiquitous on, but you can also get paid on products that benefit as a result."[49]

For his part, Clark's overarching imperative remained speed: speed of development, and speed to market. Clark was impatient, but he also believed that this was a once-in-a-lifetime market opportunity—if they could only get big enough fast enough, problems like "making money" would take care of themselves. But they had to be first to market—or at least, second to market. Mosaic was still a glorified research project that could be usurped by a more polished product. At least, that was what Mosaic was for the time being.

■

THEY WERE RIGHT TO HURRY. In May 1994, the original NSCA Mosaic browser code was licensed to a company named Spyglass, Inc., which had been formed to commercialize NCSA technology. It turned out that by poaching their student workforce, Clark and Andreessen had awoken the NSCA to the financial value of the web browser as a product. Spyglass would use the NCSA's technology to begin a lucrative business creating browsers and licensing them to various outside companies.

At around the same time, the University of Illinois threatened to sue on the NCSA's behalf, claiming that the new browser was being built using Mosaic's original code. It also hadn't escaped the university's notice that Clark and Andreessen's company had originally called itself *Mosaic* Communications. In a preliminary attempt to appease the university and avoid litigation, the name of the company was changed to Netscape, and the programmers submitted to what amounted to a forensic auditing of their work, despite the fact, as Jon Mittelhauser says, "We didn't *want* to take any of [the old Mosaic] code, that's the thing! We *wanted* to start from scratch. We wanted to do it right."[50]

As this was going on, on October 12, 1994, the marathon sessions of hard work in Mountain View paid off. A beta version of the new web browser, version 0.9 of a program eventually called Netscape Navigator, was made available on the web for anyone to download at midnight.

"When we announced it on [the WWW-Talk message boards, the same place the Mosaic browser had been launched] we had a different sound effect for different downloads," Aleks Totic recalled. "We're all sitting in this room, listening for the sounds and as soon as the email goes out there's some guy in Australia trying to download it and you hear the smashing glass. Then a couple of minutes of silence. And then a cannon. And it started getting faster and faster. We were all just sitting there drinking beer and coding a little bit and listening. And within like five or six hours there was just a cacophony of explosions and croaks and lightning and cannons. Because people were downloading it from all over the world and we're like, 'OK. We've got something.' Everybody loved it."[51]

Netscape Navigator was a generational improvement over the other browsers then available. Navigator was fast, even working under the constraints of the slow modem speeds that were standard at the time. By some measurements, Navigator could load a webpage ten times faster than Mosaic. Early reviews from users and from the media were rapturous. *Businessweek* said that Navigator could "make the Internet a mass medium for home shopping, banking and a host of other services."[52]

Over the next few months, beta versions, and then the official 1.0 version, were downloaded about 6 million times.[53] The Navigator browser quickly gained a reputation for being fast, stable, and feature-rich. It included so many web innovations that weren't supported by existing browsers that a unique new phenomenon began. Website after website on the still immature web started posting little buttons that read "Best viewed in Netscape Navigator" with a link that sent you to the download page. Just as had happened with Mosaic, webmasters and web creators wanted to show off the cool new things that Navigator allowed them to do, so they steered their users to the new browser organically.

It was estimated that 20 million people were on the Internet at the time of the beta release of Navigator. This represented amazing growth in the eighteen months since Mosaic's own beta release. In what felt like no time, Navigator quickly eclipsed Mosaic: at the start of 1994, the original Mosaic and its variants controlled 95% of the web browser market. By the end of October, a mere two weeks after release of the beta version, Navigator had captured 18% of the market, and by early 1995, Navigator was used by 55% of web surfers. By 1996, 45 million copies of Navigator had been downloaded, representing a full 80% of the browser market.[54] By that point, Mosaic's share of the browser market had shrunk to a mere 5%.[55]

"Now people take for granted that they'll put out a version of something and a million copies will be downloaded in a week," Netscape employee John Giannandrea said. "But nothing like that had ever happened before."[56] As John Naughton said in his *A Brief History of the Future: From Radio Days to Internet Years in a Lifetime*, "Netscape had effectively launched an era when you could finish a product one day and have hundreds of thousands of users the next. The old era of two-year product cycles was over."[57]

Indeed, no sooner was Navigator 1.0 out the door than the team started work on version 2.0. The product launches couldn't come soon enough. The company had burned through much of the $13 million that Clark and Kleiner Perkins had invested thus far. But the cash-flow issue would be resolved by the official arrival of Jim Barksdale as CEO.

Barksdale brought old-school business acumen to the young company. In the few short months that their main product left beta and existed in the marketplace for the first time, Netscape was on track to do $3 million in revenue for the first quarter of 1995. But Barksdale quickly discovered that he could do better. Early in his tenure, he sat down with Bill Kellinger, who ran the sales department. At that point, the sales team consisted of three overworked phone representatives who were handling more than a thousand calls a day. When Kellinger showed Barksdale these call-volume numbers, the new CEO was aghast. In effect, Netscape was turning away paying customers because there weren't enough people to answer the phones. "If I give you more people," Barksdale asked Kellinger, "how much more revenue can you do?" Kellinger figured that if he put another three people on the phones, making a total of six, he could triple Netscape's revenue. "You mean you can do nine million dollars in the second quarter?" Barksdale asked incredulously.[58]

Kellinger got his extra phone reps. Second-quarter sales reached nearly $12 million.

Where were these sales calls coming from? Well, corporate America, just as Marc Andreessen had hoped. The "sort of free" strategy backed up by official licenses was paying off.

"We could look at our server logs and we could tell who was coming in and using the browser," says Jon Mittelhauser. The sales and marketing team examined those logs and would say, "'Oh, Oracle has 20,000 people using.' Call up the IT guy at Oracle and say, 'You've got 20,000 unlicensed copies, you owe us X dollars.' We were making millions of dollars off of browsers."[59] Browsers that were ostensibly free.

By the end of 1995, Netscape would collect approximately $45 million in browser revenue alone.[60] This growth forced the young company's human resources department into overdrive, as the head count topped 250 by the summer of 1995. It would double that number by the end of the year. Based on the impressive growth statistics, CEO Barksdale was able to rustle up a second, $17 million investment round that included publishing companies Knight Ridder, Hearst and Times Mirror, as well as the cable company TCI. Netscape was valued at $150 million. Barksdale also put the legal issues with the University of Illinois to bed by settling out of court. Netscape agreed to pay the university $2.2 million in damages, with an additional payment of $1.4 million depending on future business deals. The university split the money with Spyglass, the Mosaic licensee. Netscape offered to pay with shares of the company in lieu of cash, but this was rebuffed. That refusal would cost the University of Illinois tens of millions of dollars when Netscape had its initial public offering.

And an IPO was definitely coming. In May 1995, Spyglass filed to go public. That was all the impetus Jim Clark needed: at the June meeting of Netscape's board of directors, he began agitating for Netscape to do its own IPO, and the sooner the better. CEO Barksdale and the chief financial officer, a former Morgan Stanley banker named Peter Currie, weren't so sure. The traditional rule of thumb was that a company didn't go public until it had three years of solid revenue growth; Netscape only had two quarters of any sort of revenue at all. It was also tradition for a company to show at least three quarters of profitability before an IPO; Netscape was on track to see profitability, but not until the end of the year. *And then* there was the small fact that the company wasn't even a year and a half old at that point. When Clark's own Silicon Graphics had gone public in 1986, it had been in business for five years!

Jim Clark wasn't concerned with any of these traditional measuring sticks. At his urging, Netscape filed papers for an initial public offering on June 23, 1995, four days before Spyglass's debut on the markets. Clark reasoned that Netscape had majority market share in a young software market that seemingly had nothing but growth in its future. A user base of more than 5 million had to have some value to Wall Street. And software companies were the darlings of Wall Street at the time. Software is a high-margin business; a hit software product can be a gold mine, and investors were eager for a new breed of startups.

Netscape was not the first company to go public without significant profits (or even revenue) to speak of. Speculative enterprises like mining, energy and pharmaceutical companies often IPO early in order to raise money on the promise of a big score sometime in the future. But Netscape was the highest-profile of a new breed of company that was looking to profit off the promise of the Internet. The splash would popularize the notion that the web and the Internet were new markets of unusual possibility and unique prospects. The web could potentially be a motherlode of a marketplace, and because of this, Internet companies would be held to different standards of valuation. In the dot-com frenzy that would follow, numerous IPO candidates could and would point to Netscape as a company that had gone public with zero revenues, only to ride the parabolic growth of the Internet to hundreds of millions in revenue in a few short years. Just as important, Netscape made it okay to go public even if you were only a few months old. Better to raise all the money you could and grab as much market share as possible before competitors could beat you to it.

Another key enabler of the Netscape IPO was the fact that Wall Street was buying into Marc Andreessen's platform strategy. The investing community believed Navigator was building a platform on the web, and therefore, Netscape could become the next

Microsoft. "A lot of people had missed out on the Microsoft IPO because they didn't believe in PCs" said Frank Quattrone, a Morgan Stanley banker who would help take Netscape public. Buying Netscape stock as soon as it IPOed was, in a lot of people's estimation, a once-in-a-lifetime chance to jump on board the Microsoft of the next technology era.[61]

Every IPO is preceded by what is called a "road show," where the principals in the firm go around the country pitching their company as an investment to stock analysts, investors, mutual funds, pension funds and the like. Netscape's road show was like the world tour of a pop star. In New York, people were turned away when a 500-person ballroom was filled to capacity. Many in the crowd showed up not to ask questions about the company, but to find out more about the Internet in general.[62]

■

THE AUGUST 1995 NETSCAPE IPO was the biggest thing Silicon Valley had seen in a while. For the first time in years, there was fire in the Valley again. Netscape seemed to have bottled it, and Wall Street was ready to buy it.

The morning of the IPO, Jim Barksdale had given strict instructions that Netscape employees should not discuss the stock price and should instead keep working like it was business as usual. When Jim Clark got into the office that morning, he noticed that his own personal assistant had ignored the injunction and put a live electronic ticker tape above her desk. Clark decided not to reprimand her (she was a shareholder too, after all).

"It was exciting," remembers Jon Mittelhauser. "We yelled and screamed and all that stuff. Then an hour later we were back to work. Because none of us really understood what was going on. And all of us had something we were in the middle of doing."[63]

For his part, Marc Andreessen wasn't even awake. He had been up until three the night before, working. When he woke up at 11 A.M. Pacific Time and logged in to Quote.com, the stock was finally trading, so he missed all of the drama of the delayed open. "Then," Andreessen remembered, "I went back to sleep."[64]

Andreessen went back to sleep a multimillionaire. A few short months later, when Netscape's stock price peaked at $171 a share, more than six times the price at the IPO, so were all of those "kids" from the NCSA basement. Their 100,000 shares apiece were worth almost $17 million, more even than the $10 million that Jim Clark had promised. By that point, Clark's own 20% equity stake in the company meant that he had achieved the billionaire status he had coveted for so long.

Netscape was Clark's second billion-dollar company, after SGI, but it wasn't just the bigwigs who were getting rich—it was the engineers and the secretaries as well. In his rush to go big and hire big, Clark had been generous with everyone he wanted to recruit. Here was the start of the cherished dot-com–era idea that all you had to do was pick the right company and get in early enough—so that even you, a lowly engineer, could make millions of dollars off of stock options. Netscape started the gold rush for everyone and everything, for engineers, for IPOs, for stocks, for cockeyed business plans. More than anything, the speed of Netscape's ascent shocked people. It had taken twelve years before you could begin to talk about all the millionaires Microsoft had minted. Netscape had done it in fifteen months. Wall Street and Silicon Valley had learned a valuable lesson; the web and the Internet in general was the Wild West, a land grab. The key was to get established first and dominate your market before competition could even notice. Andreessen's platform strategy seemed to be a proven concept. Early profits were not important at all. Revenue was important, but not entirely a requirement. The more valuable thing was to show a sense of "Netscape speed," the ability to be nimble and a willingness to chase markets and market share; to sense your moment of opportunity and be willing to go after it immediately. As the journalist and author Michael Lewis described it later, "You had to show that you were the company not of the present, but of the future. The most appealing companies became those in a state of pure possibility."[65]

Netscape had made entrepreneurship cool in America for the first time in a long time. For most of the previous few decades, kids aspired to be rock stars, athletes, or maybe astronauts (and stockbrokers, briefly during the 1980s). But few people had thought that starting a business could turn you into a rock star, let alone provide a decent path to fabulous wealth. All of a sudden, here was a high-profile Cinderella story wherein a bunch of college kids had taken a chance, gotten rich, and become—to the financial press at least—famous. Andreessen's *Time* cover story ran on February 19, 1996. Fourteen years earlier, on February 15, 1982, the *Time* cover boy had been a twenty-six-year-old tech superstar named Steve Jobs. The headline in 1982 read: "Striking It Rich." It signaled to the world that the first Silicon Valley revolution/gold rush was in full swing. In 1996, Andreessen was pictured barefoot and snarling (or yawning, depending on your interpretation) sitting under the headline: "The Golden Geeks." For those who were listening, and for those of a certain technological persuasion (and perhaps for those of a certain age), the message was loud and clear: a new revolution was on, and a new gold rush. Netscape laid the groundwork for the cult of the entrepreneur that is still with us today, and an entire generation took notice.

Significantly, the Netscape story wasn't all hype. In eighteen months, Navigator reached an installed base of 38 million users.[66] From $17 million in revenue at the time of the IPO, Netscape would surge to $346 million in sales the very next year, 1996, and $533 million in 1997.[67] In three years, Netscape grew revenue to levels that it had taken Microsoft almost fourteen years to reach. By some measures, Netscape was one of the fastest-growing companies in history. People believed that Netscape *would* become the next Microsoft, the colossus of the new, coming Internet Era. A browser like Navigator would be the Internet's operating system, replacing the old PC operating systems like Windows. Navigator 2.0, appearing shortly after the IPO, integrated email and newsgroup features, and added support for plugins, which allowed third parties to integrate ever more sophisticated features. Navigator was now, as one Netscape product manager put it, "a real platform that people could actually write applications to."[68] It was Microsoft's tried-and-true platform strategy, but for this whole, new, limitless digital realm. The press was already calling Marc Andreessen "the new Bill Gates."

There was just one problem. Why would Bill Gates willingly cede his throne atop the technology industry? Why would he allow his platform to be supplanted by a new one without a fight? In fact, one final but key reason Netscape had raced headlong toward an IPO was that Netscape management was terrified of Microsoft. They knew Netscape had to get as big as it could and gain as much market share as possible before Bill Gates and Microsoft woke up to the Internet in general and the potential of the web browser market in particular. In his autobiography, Clark likened Gates to the evil Lord Sauron from *Lord of the Rings*, "whose all-seeing eye searched ceaselessly for any threat to his tyranny." As the business press beat the drum that Netscape could be the new Microsoft, Gates couldn't help but hear of this new threat to Microsoft's then-total hegemony. And if he had somehow missed all of these messages, Gates had to have heard a particular, infamous jab from Marc Andreessen himself. A few weeks after the IPO, Andreessen was quoted in *InfoWorld* magazine saying that Netscape would turn Windows into "a mundane collection of not entirely debugged device drivers."[69]

2

BILL GATES "GETS" THE INTERNET

Microsoft and Internet Explorer

Netscape was right to fear Microsoft. These days, it's almost impossible to imagine how completely Microsoft dominated the computer industry at the dawn of the Internet Era. Bill Gates's company had been founded right at the dawn of the personal computer revolution. Like other pioneers of the PC era, Gates had a vision for a computer ecosystem of billions of machines, and all he wanted was for every one of those machines to have his software on them. Microsoft's corporate motto was, famously, "A computer on every desk and in every home." Early employees say that the original motto (before the lawyers advised Microsoft to tone it down) was: "A computer on every desk and in every home, *running Microsoft software*."

By the early to mid-nineties, Microsoft's operating systems were on 70% to 90% of the computers sold around the world. This dominance meant that by 1994, Microsoft could boast a $38.5 billion market cap; its market valuation would soon surpass longtime tech-industry standard-bearer IBM.[1] In the previous five years, Microsoft's annual profits, revenues and stock price all quadrupled.[2]

At least at first—and to Netscape's great relief—Bill Gates was not even remotely paying attention to the Internet. Almost all of Microsoft's resources were at that point being marshaled toward the development of a program codenamed "Chicago," the greatest update to Microsoft's operating system to date. Better known as Windows 95, this release would represent the absolute pinnacle of Microsoft's primacy in the tech industry.

If you had asked Bill Gates in 1994 if Microsoft was prepared for the next wave of computing, he would have said yes: that next wave would be named Windows 95. But if you pressed him further and asked about a different *kind* of computing, about something more networked and interactive—about something, in short, like what the Internet would become—he would have said, "Absolutely." But he wouldn't have used the term "Internet" to describe the future as he saw it. He might have mentioned a personal favorite acronym, IAYF (Information at Your Fingertips) or used a term like "information superhighway." As far as he was concerned, Microsoft already had that locked up as well.

■

IF YOU WERE ALIVE in the early 1990s, chances are you remember the term "information superhighway." It was bandied about in all corners of the media. It was the Jetsons-like futuristic media technology that many in various industries were convinced would change the world. You could be forgiven for assuming that the information superhighway *is* the Internet, or at least, the Internet is what the information superhighway became. But that is wrong.

The information superhighway was the fever dream of the telephone industry and the cable industry and the computer industry and even of Hollywood. The idea was that we'd all be linked together via a Frankenstein-like combination of the television and the PC. We'd be able to shop from home, and exchange video chats with each other, and rent movies on demand and receive personalized news and media based on our interests. I know. Sounds exactly like the Internet we know today. But all of this was supposed to happen on your television.

TVs were going to become interactive. More than a decade before our phones got "smart," the tech gurus and the big-money guys were convinced that televisions would become "smart" and *that* would be the innovation that would really change everything. The colossus of the cable industry, John Malone, announced a future of five hundred channels, shopping and movies on demand. Media titans like Time Warner's Gerald Levin predicted: "Once you digitize the material, then the consumer can summon the material at will. It's profound: not the technology but the psychology."[3] Raymond Smith, CEO of Bell South, opined, "The three principal consumer communication devices—computer, TV and telephone—are margining into one, and as they do, so too are the distinctions among once-separate business."[4] On April 12, 1993, a special issue of *Time* magazine headlined: "The Info Highway: Bringing a Revolution in Entertainment, News, and Communication."

Why was everyone so sure that television was going to be the medium that delivered interactivity to the mainstream? When Smith was asked this by *Wired* magazine, he replied: "Because that's where the people are. You've got to start with entertainment," Smith said. He simply could not envision that computer networks would be able to deliver this anytime soon. And even if they could, "you're not going to watch television on a little monitor. You're going to watch it on a big set. That's what you'll use when you want entertainment, and you'll use the PC and keyboard when text is more important."[5]

To a large degree, Bill Gates shared this vision. He came from the world of computers, but even to him, computers were still hopelessly nerdy. Television was decidedly mainstream, technologically sophisticated and, crucially, high bandwidth. Gates believed that the networked future would come via the TV because that was where the bandwidth was; 14.4 modems, clunky dial-up connections—these could not deliver the multimedia extravaganza Gates was envisioning. But high-bandwidth coaxial cable (or maybe DSL lines from the telecom companies; or maybe satellites)—could do the trick. Gates shared the vision of an interactive, smart-television world. In industry circles, Gates began to evangelize IAYF heavily as the future of all of these overlapping industries. He agreed that the living room was the logical place for this to happen. That's where the eyeballs were and that's where the existing infrastructure was.

Throughout the early 1990s, Gates took meetings with all and sundry, from film studio moguls to telecom executives. All of this was in aid of one common goal: making sure that no matter what the telcos, cable companies and Hollywood studios had planned, Microsoft would be a part of it. It was a repeat of the playbook that had won in computing: Bill Gates just wanted his software in every device that took up pride of place in the living room.

Gates was not alone in chasing this interactive television dream. If you read business and technology magazines from the period, all the way through the summer of 1995, the articles were all about the information superhighway, the convergence of telephony, television and computing, and which corporate conglomerate would come out on top. All around the country, hundreds of millions of dollars were poured into interactive-television initiatives. The biggest project, and the one to get the most attention, was Time Warner's Full Service Network in Orlando, launched to 4,000 homes in January 1995.[6] It was made possible via hardware from Jim Clark's Silicon Graphics, which helped build the set-top boxes. The service had movies on demand, interactive video games, print content from Time Warner's stable of magazines, and a virtual shopping mall where couch potatoes could order items from the Sharper Image, Crate and Barrel, the U.S. Postal Service, a Dodge dealership and a local supermarket.

"I challenge anybody to say that video-on-demand isn't what the consumer wants,"[7] Jerry Levin, CEO of Time Warner, declared. He could have just asked the consumer. One by one, all of the interactive TV experiments failed spectacularly. A GTE test in El Cerrito, California, was designed for 7,300 households. Only 350 ever signed on.[8] The bestselling item in the vaunted Full Service Network virtual mall? Not new cars or groceries, but postage stamps.

The "interactive TV" aspect of the information superhighway was largely a bust. But this didn't concern Bill Gates too much. He didn't care who won the mad scramble to deliver this golden future: cable, telephone, satellite or other. Microsoft would sit back and let others lay the groundwork and infrastructure of a fully connected IAYF world. Once all the kinks were worked out, Microsoft would swoop in and overlay its next-era platform on top of everything and take a generous cut for doing so. It was a strategy that had worked for Microsoft time and again in the 1980s: let others do the hard work of proving a market, then come in and dominate it once the dust had settled. Various industry estimates said that true broadband wouldn't become common in North America until the turn of the century (an accurate prediction, as it turned out.) Gates believed he had time to wait. The networked world he was envisioning couldn't happen until broadband was ubiquitous. The future wouldn't happen overnight.

▪

EXCEPT, OF COURSE, IT HAD.

It all came down to this: no one in tech, no one in media, no one from Bill Gates to Jerry Levin to Hollywood ubermogul Barry Diller had realized what Marc Andreessen and Jim Clark had realized: *the information superhighway was already here*. The Internet and the World Wide Web *were* the information superhighway. The revolution was now, and it was being delivered not by the television, but by the computer.

Part of this misjudgment was probably just generational bias. Bill Gates (born 1955), Barry Diller (born 1942), Jerry Levin (born 1939), John Malone (1941), and all the rest were baby boomers or near-boomers. They had grown up in the Age of Television. For these men, it was taken for granted that television was the apotheosis of mainstream technology, the cultural force that united all of late-twentieth-century society. Like any good computer hacker, Bill Gates had used the Internet in the 1970s and '80s. In fact, when Gates developed Microsoft's first-ever software product (a version of BASIC to be used on the Altair, the first personal computer), he had used FTP on Harvard University's computers to beam his work for storage on Carnegie-Mellon's computers. But to Gates, the Internet was like Unix: it was a technology for geeks. What average computer user could be bothered to figure out something arcane like FTP?

The Internet was not for mainstream users, as far as Bill Gates was concerned. Microsoft was a company that thrived by selling carefully controlled user experiences. Microsoft had come to prominence by making computing more mainstream and user-friendly. That was why Gates's vision for the information superhighway developed by Microsoft and its big media partners would be a safe and controlled technology, palatable to mainstream users, and above all, managed.

What Gates missed most crucially was how the latest iteration of the Internet, the World Wide Web, was different. It was, in fact, more user-friendly, and more robust than anyone realized at the time. Gates simply missed that the Internet had undergone the equivalent to the personal computer/GUI revolution that Microsoft itself had delivered in computing. The web could deliver on all of the promises of the information superhighway, and it delivered on those promises in the democratic, utopian way that so enthused early adopters of the web like Marc Andreessen. The information superhighway was interactive, sure. It let you talk back to your TV. But it didn't allow you to *create* your own television program. The web, by contrast, allowed users to consume content, and create it. Any user. Anywhere. Any kind of content. And anyone could do so outside the control of a major media corporation or gatekeepers like the cable companies or Microsoft.

A young Microsoft executive named Brad Silverberg, who joined the company in 1990, put it this way, "If you're Microsoft in the middle of 1995, the world is pretty good! You're king of the hill! The technology world revolves around you! Why would you ever want the world to change? Understandably, you don't."[9]

But the world had changed, and it took Gates a little while to understand this. The best illustration of this comes from the book that Bill Gates agreed to write sometime in the early 1990s called *The Road Ahead*. It outlined Gates's own vision of the future of technology. Published in November 1995, the index of the hardcover edition had 68 references to the term "information highway," 46 references to the term "Internet" and 4 references to the World Wide Web.

About a year later, the paperback version was released, and it had been heavily rewritten. In the paperback version of the book, "information highway" got only 39 references. The Internet, conversely, got 169. The web suddenly had 59 mentions. Why the change? Between the hardcover and paperback editions, Netscape happened.

"The Internet was the information highway everyone was looking for," Netscape's Jim Barksdale said. "They just hadn't recognized it."[10]

■

BUT THERE *WERE* PEOPLE at Microsoft who recognized it. They were younger Microsofties, a bit older than the Marc Andreessens of the world, but generally of the same Gen X age cohort. These younger executives and engineers, in various ways, and in sporadic, uncoordinated efforts of agitation, would begin a slow but steady drumbeat to wake Microsoft up to the web revolution. They did so in, probably, the only way that change can be made in large corporate environments: via the quiet, measured insurrection of memorandums.

James Allard, born in 1969, took it upon himself to become Microsoft's first intermediary to the net/web revolution. Before this time, Microsoft made little contribution to the development of the web and the Internet at large. Microsoft had no seats on standards committees. It had no one participating in the WWW-Talk forums. Allard began representing Microsoft at early Internet confabs, like the Internet Engineering Task Force, and made sure that Microsoft became a founding member of the Internet Society. In early 1993, Allard started an in-house Microsoft discussion group on the Internet called "inetdisc."[11] Out of 14,400 Microsoft employees at the time, 5 people joined. Undeterred, Allard printed a batch of Microsoft business cards that read JAMES ALLARD, PROGRAM MANAGER, TCP/IP TECHNOLOGIES.[12]

On January 25, 1994, around the time that Marc Andreessen

was first getting to know Jim Clark, Allard wrote an internal Microsoft memo titled "Windows: The Next Killer Application on the Internet." The memo outlined the recent explosion of growth on the Internet and of Mosaic. Allard asserted that the Internet represented a great opportunity for Microsoft. "By embracing current technologies available on the Internet," Allard proposed, "we position Windows as the choice system for interactive Internet services and prepare for the shift to the native IAYF (Information At Your Fingertips) technologies offered [by Microsoft Products]."[13]

One of the people cc'ed on the memo was Steven Sinofsky. Sinofsky was another young Microsoftie enamored with the Internet. As the technical assistant to the CEO, Sinofsky's purview was to keep Bill Gates abreast of industry and technology trends. Also a Gen Xer and heavy Internet user in his college days, Sinofsky had given Gates a personal tutorial on an array of Internet tools as recently as October 1993, including a browsing session on the nascent World Wide Web. At the time, Gates was intrigued but not overly impressed.

At the time of Allard's memo, Sinofsky had taken part in a recruiting trip to his alma mater, Cornell. In between interviewing bright young prospects for possible employment with Microsoft, he couldn't help but notice how prevalent the Internet had become in everyday campus life. At least among these college kids, things like email, web browsers and newsgroups weren't opaque, fringe technologies. Seemingly overnight, they had become mainstream. On Valentine's Day, February 14, 1994, Sinofsky wrote a memo of his own, with the title "Cornell is WIRED!"

Around this same time, Microsoft began hedging its bets when it came to trends in networked computing. Microsoft studied the existing consumer online services like Prodigy, CompuServe and America Online. These services had nothing to do with the Internet or the web (more on that in the next chapter) but they were training a small yet growing population of pioneers to begin to settle cyberspace. Microsoft began development of an online service of its own, a service that would eventually be known as the Microsoft Network, or MSN. It was slated to launch alongside Windows 95.

At a strategic retreat for upper-level Microsoft management on

April 7, 1994 (two days before Netscape was officially founded), Gates began to entertain the possibilities of the Internet in a more serious way. "Everywhere I go, people ask me about how Microsoft will be on the Internet,"[14] Gates said to launch the retreat. But did this mean simply enabling Internet tools within the forthcoming Windows 95? To Gates's mind, the biggest question of all was how Microsoft could make money on the Internet. Seemingly everything on the Internet was free. This was not a small point to overlook. Gates could see how Microsoft could make money on the information superhighway, by serving as the gatekeeper and toll collector. But the freeware, ungoverned, unsettled Internet didn't seem to offer a similar opportunity.

Allard and Sinofsky were ready to argue these points. Sinofsky had put together a comprehensive 300-page catalog of Internet items he had collected, designed to show the breadth of what was already out there.[15] These curios included sites that were beginning to host not just images but also streaming and downloadable music and video. Allard followed up by evangelizing for incorporating the Internet into everything Microsoft would be doing for Windows 95.

Two weeks after the executive retreat, Gates issued a memo summarizing key talking points. Gates wrote: "We want to, and will, invest resources to be a leader [in] Internet support, fully understanding that if we are wrong about this it will have been a mistake."[16] But at least they would be covering their bases.

In short order, a couple of related events would further evolve Gates's thinking. As a part of dipping its toe in the Internet waters, the idea of a Microsoft web browser was discussed in earnest, spearheaded by a young Microsoft engineer named Ben Slivka. In August 1994, Slivka began "cataloging" key Mosaic interface features as a way of determining the basics Microsoft would need to master in order to launch a competitive browser.[17] At the same time, Microsoft started shopping around for existing solutions and entered talks with a small software company called BookLink Technologies, which had a Windows-based browser called Internetworks. Suddenly, in November 1994, BookLink announced that the entire company had been acquired for $30 million. The buyer was none other than America Online, the online service that MSN was intending to supplant.

Thirty million dollars for a browser? "That woke us up," said Brad Silverberg, one of the executives in charge of Windows 95 development. "We had to be a lot more aggressive, a lot more lively. Time was ticking faster in this new world."[18]

Moving to plan B, Microsoft tentatively reached out to Netscape to learn about their Navigator browser; maybe that could be licensed for Windows 95. Here Microsoft received another shock. Netscape rebuffed Microsoft's overtures completely, and somewhat rudely. Netscape did not have any intention of doing business with Microsoft.

Who were these Netscape guys and what did they have against Microsoft? Why weren't they willing to do business? It was puzzling.

And then of course came the release of Netscape Navigator itself. Suddenly, all the pieces fell into place. With the launch of Navigator came the millions and millions of downloads and all the attendant media attention. As *Fast Company* put it, "Virtually overnight, Netscape was perceived as the defining company of the Age of the Web."[19] Much of the related hype Netscape received came with those pointed barbs that seemed to be aimed squarely at Microsoft. All those headlines suggesting Marc Andreessen as the next Bill Gates? That couldn't help but turn Gates's head.

Nothing got under Gates's skin like discovering a software market he did not have dominant control of. Netscape had proven that web browsers were an enormous market. Furthermore, lots of people inside and outside of Netscape were already seeing what Marc Andreessen had seen: the browser could be a software platform capable of supplanting traditional operating systems like Windows. If, in the future, people could live their lives and do their work entirely online, then what would be the need for a desktop OS?

Yet another memo, this time from Slivka, still agitating for the browser project. Slivka's missive cut right to the greatest threat that the Internet posed to Bill Gates's vaunted business model. Its title read simply "The Web Is the Next Platform."

On May 26, 1995, Gates wrote his own memo to senior Microsoft executives, entitled "The Internet Tidal Wave." It would become one of the most famous documents of the Internet Era. In it Gates announced that the number-one priority for Microsoft, in every facet of its business, was now the Internet. Every product manager should stop what they were previously doing and start considering how the Internet could affect their products, or how their products could make an impact on the Internet.

Gates was not afraid to acknowledge his past reticence. But he made clear those days were over:

> I have gone through several stages of increasing my views of its importance. Now I assign the Internet the highest level of importance. In this memo I want to make clear that our focus on the Internet is crucial to every part of our business. The Internet is the most important single development to come along since the IBM PC was introduced in 1981.

And Gates made clear who the first target would be as Microsoft now changed direction.

> A new competitor "born" on the Internet is Netscape. Their browser is dominant, with 70% usage share, allowing them to determine which network extensions will catch on. They are pursuing a multi-platform strategy. . . . We have to match and beat their offerings.

Microsoft would jump on the Internet in a big way, and Netscape was enemy number one. Many of the young guns inside the company who had been banging on the Internet drum for a while, wondered if it might be too little, too late. "It kind of felt like, it's great that Bill is now finally lending support to the Internet," recalled Brad Silverberg. "But at the same time it felt like he was the last executive in the company to come around."[20] Better late than never, Internet capabilities were hastily added to the already delayed Windows 95. An extra $1.5 billion was set aside for web research and development.[21] And the crash program to develop a Microsoft web browser, the key goal of Slivka's agitation, was given the highest priority.

But with this browser project, Microsoft would have to confront, both culturally and structurally, the ways that Netscape and "Internet Time" had changed the rules of the game. Microsoft was very much used to the old methods of multiyear product development schedules. Development of what would become Windows 95 had begun way back in 1991. The program was originally to be called Windows 93, in fact. To be sure, a full operating system was a more complicated thing to develop than a web browser, but Microsoft was notorious for spending four years on a project with multiple delays. This sort of thing simply wouldn't fly if Microsoft had any hope of challenging Netscape in the browser market.

So, Microsoft did what it had to do: it cut corners. Having lost BookLink to AOL and having been rebuffed so arrogantly by Netscape, Microsoft was forced to turn to the most logical remaining choice: Spyglass, Inc., the company approved by the University of Illinois to commercialize the original Mosaic web browser. Microsoft signed a $2 million licensing agreement with Spyglass to use Mosaic code for Windows 95. Irony of ironies, the code that would be the basis for Microsoft's web browser (and the weapon Microsoft would soon wield against Netscape) was a descendant of the same code written by Marc Andreessen and Eric Bina a few years before at the NCSA.

The original Internet Explorer team was a commando unit of five or six programmers, including Slivka, and led by Silverberg. Their orders were to get the browser done, quick and dirty if necessary. They would follow the traditional Microsoft game plan: the first version would be a copycat product that didn't have to be great; it just had to be good enough. Subsequent versions would be better. "We needed to get something into market quickly as a placeholder," Silverberg recalled later.[22] Once they put their stake in the ground, Microsoft would revert to form and throw everything it had at the problem until a Microsoft browser could be truly competitive.

Bill Gates had one more favorite trick up his sleeve to level the playing field quickly. On its release in August 1995, Microsoft announced that Internet Explorer would be free. Not kinda-sorta free, wink-wink free, like Navigator was. But 100% free to anyone and everyone, even corporate users. As Gates himself admitted, "One thing to remember about Microsoft, we don't need to make any revenue from Internet software."[23] The intention was to bundle Internet Explorer as a component of Windows 95. Microsoft wanted users to think of Internet Explorer as a core function of Windows. It would be a routine part of the OS, just like screen savers or disc compression utilities or file managers. Internet Explorer would sit prominently on every Windows machine, a smiling blue "e" icon on every desktop that ran Windows.

This was not a small consideration. When Windows 95 finally launched on August 24, 1995 (two weeks after the Netscape IPO), it was possibly the largest product launch in history. Computer stores around the world opened at midnight and lines of eager customers queued up to be the first to nab a copy of the program. Comedian Jay Leno joined Bill Gates onstage to emcee the official launch event. ("To give you an idea of how powerful Windows 95 is," Leno joked, "it is able to keep track of all of O.J.'s alibis at once.")[24] In New York, the Empire State Building was lit up in the colors of the Windows 95 logo. And famously, the Rolling Stones were paid a reported $14 million for the use of their song "Start Me Up" in Windows 95 commercials. All in all, Microsoft spent around $300 million making sure that Windows 95 was a blockbuster.

Having Internet Explorer piggyback on Windows 95 was therefore a powerful strategic move. The Internet was still very young, and plenty of users would encounter it for the first time via Windows 95. The first versions of Internet Explorer were not very well reviewed, and compared poorly to Netscape Navigator when it came to features and performance. But Internet Explorer was right there automatically on every Windows machine. To get a copy of Navigator, conversely, you had to search it out and download and install it yourself—not an easy feat for Internet newbies.

After joining battle with Netscape, Microsoft copied its foe and began to iterate relentlessly. Versions 2 and 3 of Internet Explorer were developed concurrently. By Internet Explorer 3.0, reviewers were beginning to say that Microsoft had at least a competitive browser. This all had a gradual but accumulative effect on Netscape. At first, Netscape Navigator's share of the browser market remained dominant, but Internet Explorer started making inroads, increasing from virtually nothing in 1995 to 20% in 1996 and 40% in 1997. There was little Netscape could do in the face of the Microsoft onslaught. Sentiment in the industry and on Wall Street began to turn. "Microsoft may still be No. 2 in the Internet race, but it's rapidly closing the gap," *PC Week* declared.[25]

Netscape's entire Get Big Fast strategy had been predicated on making the Navigator browser the de facto standard before competitors like Microsoft noticed. The hope was that they could achieve a market share and a mind share that would be impossible to dislodge. But within eighteen months of setting off the big bang that announced the coming of the Internet Era, it looked as though even the head start Netscape had managed to earn might not be enough to fend off Microsoft's muscle. "People aren't asking anymore if Microsoft will be killed by the Internet but whether Microsoft will dominate the Internet," a market researcher from Gartner Group told *Newsweek*.[26] Steve Jobs told *Wired* in 1996, "If you don't cross the finish line [if competitors couldn't outmaneuver Microsoft] in the next two years, Microsoft will own the Web. And that will be the end of it."[27]

3

AMERICA, ONLINE

AOL and the Early Online Services

The way that Microsoft leveraged its platform was not exactly subtle. If you bought a computer in 1995, there was a 90% likelihood you purchased a computer with a Microsoft operating system preloaded on it. If you were a web neophyte in 1995 and you wanted to give this "web thing" a try, chances were very good that you clicked on the bright blue Internet Explorer "e" icon to do so.

Netscape recognized the value of the preloaded icon on a Windows 95 desktop and tried to cut deals to get Navigator preloaded on various computers. Compaq was one such manufacturer that began replacing Internet Explorer with Navigator on some of the models it sold; or, at least, it offered consumers a choice of preloaded browsers. But in June 1996, Compaq received a "Notice of Intent to Terminate" from the Microsoft legal team. In no uncertain terms, Microsoft threatened to cancel Compaq's Windows 95 license unless the company returned the Internet Explorer icon to the Windows 95 desktop on all computers it shipped.

Compaq, of course, backed down.

Netscape wasn't the only major player in 1995–96 that felt disadvantaged by Microsoft's, shall we say, political decisions vis-à-vis the desktop real estate on Windows 95. Before a user could even select a browser to surf the web with, she first needed to engage a service that would allow her to "log on" to the Internet. She needed an Internet service provider, or ISP. It just so happened that Microsoft provided a strategic default solution for that as well: the Microsoft Network.

MSN had been developed to compete with existing online services such as Prodigy, CompuServe, and especially AOL. But along with the great pivot toward Internet Religion, MSN had been quickly reconfigured as an online service–ISP hybrid. Then on March 12, 1996, a curious thing happened. AOL—that online service competitor that MSN had been designed to vanquish—announced that it would make Internet Explorer the default web browser for its service. No money changed hands, but as part of the "partnership," an AOL icon would be placed in a new folder on all Windows desktops called "Online Services."

The quid pro quo was implicit, if not explicit. Microsoft would grant AOL the desktop real estate it was fighting tooth and nail to deny to Netscape. It turned out that Microsoft saw the battle for the browser as the key strategic war it was fighting in this dawning Internet Era. Bill Gates felt it was imperative to grow Internet Explorer's market share and surpass Netscape's Navigator. He made this decision despite the fact that Microsoft had already spent hundreds of millions of dollars to develop and market MSN. Gates doubled down on software at the expense of online services. He made the calculation that the browser wars were more important to win than the scramble to connect people to the Internet. Subsequently, Microsoft struck similar deals with CompuServe, the number-two online service, as well as AT&T's Worldnet Internet service and NETCOM, a leading independent ISP. All of these new partners got icons in the "Online Services" folder. Netscape Navigator remained something you had to download yourself.

In the coming years, MSN would always be seen as an also-ran behind the eventual online service leader, AOL. Later in the nineties, as the browser wars faded into memory—and especially near the turn of the century, as America Online grew to become the one truly dominant player on the Internet that had the muscle to go toe-to-toe with mighty Microsoft—many in the industry would wonder if Gates might have picked the wrong strategic horse to champion.

■

ONLINE SERVICES HAD a long history that predated the World Wide Web. In the 1980s, when PCs were still struggling to find a "killer" use case that would justify their entry into Americans' lives, online services were dreamed up as "something else to do" with computers once you brought them home. Online services promised games, unique content from trusted media properties, software downloads, databases and vague concepts of real-world utility like online banking. PC manufacturers started bundling these services with their machines as an extra selling point to entice consumers. The fact that consumers would have to cough up hourly fees to "dial in" and use these services (fees that were shared with the PC manufacturers) didn't hurt either.

The granddaddy of the online services was CompuServe, which was born in 1969 as the Compu-Serv Network. Compu-Serv began life as a time-sharing computer service, allowing businesses to rent computing time from remote mainframes during business hours. The tax preparation company H&R Block purchased the company in 1980, and the focus on consumer online services expanded. Renamed CompuServe, the service developed a suite of prepackaged features like newsfeeds, databases and one of the world's first online chat applications, called the CB Simulator. These features became the basic template for what an online service could provide users. Consequently, CompuServe became the home of many online firsts. The first recorded online wedding took place in 1983 between two users who met on the CB Simulator and thought it fitting to say their vows in the medium that brought them together.[1] CompuServe became the first online service to offer Internet connectivity in 1989, when it allowed its proprietary email service to send messages to outside email accounts. CompuServe also pioneered online commerce with what it dubbed an "Electronic Mall." Even the humble

.gif graphics file format, still popular on the web today, was developed in-house at CompuServe. But the main feature of CompuServe throughout its life was its forums, hundreds of moderated special-interest sites catering to almost every interest and niche imaginable. CompuServe gained a reputation as the geek and hobbyist's playground, with forums catering to everything from stamp collecting to *Star Trek*.

Other companies copied CompuServe's model, launching with a varying mix of email, forums, bulletin boards, software libraries for download, and chat. They all had one thing in common: they assumed that a user would be somewhat computer-savvy. Another early online service, Prodigy, assumed the exact opposite. It was designed from the very beginning to attract mainstream users. Formed in 1984 as a joint venture between IBM and Sears (another partner, CBS, dropped out in 1986), Prodigy launched in September of 1990 on the back of a nationwide advertising blitz. Prodigy had vector-based graphics, which were primitive and cartoon-like, but were interesting and colorful compared to CompuServe's all-text environment. Newspapers and magazines repurposed some of their content for Prodigy, and big-name media personalities such as Howard Cosell and Liz Smith wrote columns exclusively for the service.

Prodigy was also conceived as an advertising medium. It was organized into magazine-like sections of interest focused on promoting or selling products. Every screen had a three-line graphic advertisement at the bottom.[2] The imprimatur of Sears and IBM attracted commerce partners such as Neiman-Marcus, Levi Strauss, Ford, Columbia Records and even Sears's archrival J. C. Penney. Prodigy hoped to make the bulk of its money via advertising fees or by taking a share of product sales.

Though the focus on ads and commerce never quite went away, Prodigy's commercial efforts quickly proved to be a bust. It turned out that when people went online what they really wanted to do was interact with each other. Prodigy's bulletin boards and email services were limited and archaic, and these systems quickly became overwhelmed. Prodigy attempted to compensate for the resulting bandwidth issues by actually *discouraging* users from using the service so much. The introduction of a 25-cent surcharge for each email a user sent over an allotted thirty emails a month led to a member revolt. Prodigy was forced to reverse course and refocus its offerings on user-created content like message boards and forums, but even then, the stodgy corporate culture of Sears/IBM was not comfortable leaving users to their own devices. "We did not think [member-to-member] communications was going to be a big part of what we were doing," Prodigy CEO Ross Glatzer told *Wired* magazine.[3] Esther Dyson, the technology analyst, summed up Prodigy's conundrum this way: "They thought they'd make revenues from people making purchases. But they discovered people were less interested in shopping on the service than communicating. And they didn't know how to charge for communications."[4]

Despite all the efforts of these pioneers, online services were still a niche business, even among computer users. By 1995, Prodigy could boast only about 1.35 million members, and that was behind CompuServe's 1.6 million accounts.[5] The company that would truly take online services mainstream was another early online pioneer that would concentrate almost religiously on allowing users to interact with each other in whatever way they wished.

■

AMERICA ONLINE ACTUALLY HAD its origins in another of the early online services, The Source, which was a competitor to CompuServe, launching in 1979. Through a convoluted series of business pivots, the company that would become AOL also shared its DNA with Control Video Corporation, a company that produced an online game service for the Atari 2600 video game console. After the video game business temporarily collapsed in the mid-eighties, the company evolved into Quantum Computer Services to produce a dedicated online service for Commodore 64 and Commodore 128 computers. It also built online services for Apple, IBM and Tandy and in 1989 evolved all these offerings into an online service called America Online, or AOL.[6]

AOL was one of the first online services to focus on Windows users, which made good business sense because it was able to ride the coattails of user adoption as Windows came into its own as the inheritor of the DOS operating system throne. This strategy also positioned the service as the most mainstream and user-friendly in the industry. AOL was built from the ground up to feature clean, dynamic modern graphics—actual pictures, not the digital line drawings of Prodigy. And first and foremost, AOL fixated on building a sense of community among its membership. AOL users were encouraged to email, argue, play and above all chat.

"From the early days, we recognized that communications—a combination of chat and e-mail—were critical building blocks," AOL CEO Steve Case would later say. "So our bias was on creating tools, empowering people, and letting them use them in any way they thought appropriate—sort of 'Let a thousand flowers bloom.' "[7]

AOL's installation process was simple. You put a disc—and later, a CD—into your computer, installed a program, clicked the icon that appeared on your desktop, and five minutes later you were online. Like CompuServe and Prodigy, the process of getting online meant using a modem to "dial in" via a phone line to an AOL computer that would serve the content to your machine. This was literally a phone call to a local number, so all the online services maintained a network of local modems for people to dial in to and avoid paying long-distance charges. While you were online, the phone line you were using was occupied, so anyone trying to call your number would get a busy signal. A monthly fee entitled users to a fixed number of usage hours per month. If a user went over the monthly limit, they were charged by the hour. On AOL, $9.95 a month got you five hours of unlimited access; each additional hour cost $2.95.[8] Once you hung up, the connection was terminated.

The sounds of first a phone number being dialed, and then the harsh crackle and hiss of the modem making a connection to the network, became a ubiquitous noise across America in the 1990s. To this sound, America Online added friendly touches: "Welcome," "You've Got Mail," and when the connection was terminated, "Goodbye." The voice was that of Elwood Edwards, a broadcaster and the operations manager of WFTY-TV in Washington, D.C., who was paid $100 for his trouble. Americans heard Edwards's friendly voice billions of times as they logged in to AOL over the course of the 1990s and early 2000s.

AOL allowed users to create screen names, or online personas that served as their identity as they surfed AOL's offerings. When you played games or posted to forums on AOL, your screen name was your calling card. Your screen name was also your email address. But most important, when you entered AOL's famous chat rooms, your screen name was your name tag.

The house of AOL was built on chat. There were public chat rooms organized by topic or theme. Then there were user-created chat rooms that were dedicated to any topic under the sun. Both of these public types of chat rooms were nominally overseen by AOL staff and/or volunteer member-monitors. It was possible to get yourself kicked out of a chat room if you misbehaved. But in addition to these, there were also private chat rooms that were invite-only and monitored by no one. In the private chat rooms, it was very much anything goes. It's a well-established notion in business theory that sex often drives the lifecycle of new technology adoption, the most famous example being the way porn movies brought VCRs into America's living rooms. It's safe to say that the popularity and growth of AOL was driven by sexy chat. Lots and lots of sexy chat.

For one thing, it was easy to attach and send photos to other users in chat rooms; trading of pornography was a common pastime. But the anonymity of the screen name meant you could be anything or anyone you wanted. Paraphrasing the famous New Yorker cartoon ("On the Internet, nobody knows you're a dog"), in AOL chat rooms, nobody knew if you were a twenty-two-year-old blonde with a pinup's body or a fifty-five-year-old divorced guy with a beer belly. Americans by the millions took to AOL chat rooms to talk dirty, role-play, and act out sexual fantasies. The company didn't like to publicize it, but chat was AOL's bread and butter. The more chat, email and picture trading users did, the more money AOL made. Some users spent hours in chat, racking up monthly overage costs running into the hundreds of dollars. An October 1996 article in *Rolling Stone* estimated that half of all AOL's chat was sexually oriented and, given the hourly fees, such adult chat netted the company $7 million a month.[9] CompuServe was too serious an operation for such lewdness, and conservative, corporate Prodigy absolutely fled screaming from any hint of unwholesome behavior on their service. By the time Prodigy started experimenting with chat rooms in earnest, AOL basically had the market cornered.

AOL has often been described as training wheels for the Internet. The nickname is apt. For millions of Americans, their aol.com address was their first experience with email, and thus, their first introduction to the myriad ways that networked computing could change their lives. Suddenly, you didn't have to exchange letters or phone calls with relatives across the country. When you wanted to say something to a distant loved one, you could just shoot them an email. And it was free! And you could attach pictures! AOL was also where people discovered communities centered around interests that heretofore had been isolated or obscure. If you were into breeding miniature dachshunds, suddenly you could connect with everyone in America who shared your interest. AOL was where Americans first wrestled with concepts of anonymity and identity in an online world. All of those dirty chatters on AOL chat rooms were at the vanguard of learning what it was like to live life in cyberspace.

In a way, AOL embodied that most American of dichotomies: wholesome, friendly, mainstream on the outside, with all sorts of prurient stuff going on behind closed doors. AOL's chief executive, Steve Case, fit at least the wholesome part of that narrative. A native of Hawaii, prone to wearing Hawaiian shirts, Case seemed like the classic middle-class baby boomer, the guy with two kids who lived next door and loved Jimmy Buffett. With his quiet, calmly earnest demeanor, Case still looked like the Procter & Gamble assistant brand manager he once was. With America Online attempting to entice users from market leaders CompuServe and Prodigy, Case put himself forward as the friendly leader of the AOL "community." Case appeared in AOL ads and would roam AOL chat rooms to personally interact with members or solve customer service issues. He sent folksy, service-wide letters to AOL users signed simply, "Steve." In the late nineties, he appeared in Gap ads modeling his trademark khakis.

Perennially the number-three online service behind the deeper pockets of Prodigy and the greater experience of CompuServe, AOL scrambled for members and struggled with mountains of red ink throughout the early 1990s. AOL was arguably the first IPO of the online era; going public on March 19, 1992, it had only done so as a part of its never-ending struggle to raise enough money to remain solvent. Shortly after its IPO, AOL could boast only 200,000 paying subscribers.[10]

Slowly but surely, however, AOL's user-friendliness paid off. Members fed up with Prodigy's heavy-handed censorship and nickel and diming over email jumped ship. And mainstream users increasingly preferred AOL's pictures and graphics over CompuServe's continued text-only environment. The previously mentioned focus on Windows users was also a major strategic coup. AOL surpassed the 500,000-subscriber mark for the first time in December of 1993.[11]

Chronically in need of fresh infusions of capital, and because it was the only independent company in the online service game, AOL had several run-ins with larger players attempting to take it over. The closest AOL came to assimilation was when Microsoft was first considering an entrance in the online services market. AOL's Windows-centric philosophy seemed like a good match, and so a Microsoft approach to AOL was made. At the very first meeting between executive teams of the two companies, Bill Gates led off by musing to Steve Case, "I can buy 20 percent of you or I can buy all of you. Or I can go into this business myself and bury you."[12] Microsoft would later assert that Gates was just thinking out loud, stating the obvious realities of the situation in a sort of philosophical manner. But that wasn't how AOL saw it. The AOL executives saw Gates's "musing" as a threat.

"We didn't trust Microsoft's motives, because we knew they could emerge as a major competitor," Case later said. "At one point in the meeting, [Russell] Siegelman [a Microsoft executive who eventually ran MSN] proposed a 50-50 joint venture, but from our point of view, it was 'OK, we'll help you build it, teach you all about it, then just when it gets interesting, you'll shoot us."[13] As another AOL executive put it, AOL was offered an unappealing choice: become "a footnote on Bill Gates' resume," or stand and fight and maybe become "the king of the online industry."[14]

AOL chose to stand and fight. It would be one of the smartest business decisions of the decade, because AOL would soon embark on a period of growth that would leave the rest of the industry in the dust.

▪

AOL'S TRIUMPH CAME in large part thanks to one of the greatest marketing campaigns in consumer history. Jan Brandt had a background in educational publishing and insurance sales before she was hired as AOL's vice president of marketing in 1993. Tasked with growing the user base, Brandt had an intuition that online services weren't a typical consumer product when it came to marketing. Selling consumers on the virtue of one online service over another was not as important as educating consumers on just what an online service *was*. It was during market research that she realized she needed to go back to basics. According to Brandt, during a focus-group study, "Someone took a computer mouse and started pointing it at the computer like a remote control. And one person put it on the floor and tried to use it like a sewing machine pedal."[15]

Brandt realized she just needed to get users to try the service. If she could somehow get the AOL experience into people's homes, the service would sell itself. Brandt approached Steve Case and requested $250,000 to mass-produce thousands of AOL trial discs to hand out to consumers for free. "It was a lot of money for us at the time," Brandt admits. But, building off her background in direct mail campaigns, she sent out her first shotgun blast of diskettes, around 200,000 pieces, in the spring and summer of 1993.[16]

The results were immediate and startling. The response rate to the first campaign was a staggering 10%, an unheard-of percentage for direct marketing. "And remember," Brandt says, "this isn't people who are saying, 'I think I want this.' These are people who are taking the disc, putting it into the computer, signing up, and giving us a credit card. When I saw that, honestly, it was better than sex."[17]

Brandt immediately doubled and then quadrupled down on the strategy. The idea was to get an AOL disc offering a free trial into the hands of every person who might conceivably get close to a computer at some point in their lives. AOL discs began arriving in Americans' mailboxes seemingly daily. Almost every computer maker shipped an AOL disc with a new computer. There were AOL discs given away with movie rentals at Blockbuster. There were AOL discs left on seats at football games. At one point, Brandt even tested whether or not discs could survive flash freezing so that she could give away AOL discs with Omaha Steaks. Once CD-ROM players became common in computers, it almost felt like there wasn't a magazine or newspaper in the country that didn't have an AOL CD inside it.

Over the next half decade, AOL would spend billions of dollars on its "carpet bombing" marketing campaign. At one point, 50% of the CDs produced worldwide had AOL logos printed on them.[18] Brandt lived in fear that competitors like Prodigy or Microsoft would copy her technique. At one point, a CompuServe executive struck up a conversation with one of Brandt's AOL colleagues at a conference. "You guys are crazy," the CompuServe exec said, referring to the CD carpet bombing and the money it had to be costing. When the AOL executive reported the conversation to Brandt, she retorted: " 'Next time someone says that, agree that I'm a dumb broad, and that you've been trying to get me fired from the company for a long time.' And the reason for that, really, was, I couldn't believe that they weren't trying it!"[19]

Prior to Brandt's marketing campaign, AOL was languishing around the 500,000-member mark. Post-Brandt campaign, AOL was signing up 70,000 new members monthly.[20] AOL passed the million-member mark in August of 1994, tripling in size in one year.[21] It hit 2 million subscribers a mere six months later and proceeded to blow past both CompuServe and Prodigy to become far and away the largest online service.[22] In May 1996, AOL surpassed 5 million subscribers, ten times the number of subscribers AOL had when Jan Brandt started shoving trial discs into packages of Omaha Steaks.

But then came Windows 95 and the MSN service that launched with it. The outlook for AOL seemed precarious. A research firm predicted that between 11 million and 19 million users would sign up for MSN in its first year, based on sales projections for Windows 95. There was a grand total of only 10 million users of online services at the time.[23] Microsoft, Case insisted, should offer all online service options as part of a level playing field. "The fact that Microsoft has an 85 percent market share . . . and wants to hardwire their own service into it in an anticompetitive way is not a good thing," he told *Wired*.[24] The AOL CEO even appeared at a joint press conference with the CEOs of CompuServe and Prodigy to release an open letter to Bill Gates, demanding the unbundling of MSN from Windows 95.

But in the end, MSN never exactly took off. Even though a reported 190,000 users signed up in the first week after MSN launched in August 1995, it had only around 375,000 users by that November. This was during a time period when AOL was bringing in 250,000 new members every month, thanks to its avalanche of free discs.[25] And then came the deal to make Internet Explorer the default browser for AOL users. After that, both AOL and the market at large knew that Microsoft's heart wasn't really in the online services business. Bill Gates had ceded de facto control of online services to AOL. If Microsoft wanted to neuter its own online offering, who was Steve Case to look a gift horse in the mouth? He threw Netscape under the bus and put AOL firmly on the road to dominance of the online services arena.

AOL would succeed in branding itself as "America, online," and neither MSN nor anyone else was ever able to challenge this. What Bill Gates didn't really appreciate at the time was how powerful being the "training wheels" for the Internet Era would eventually become.

■

JUST AT THE MOMENT that AOL was successfully fending off MSN, it faced perhaps an even greater existential threat. The web was something that online users were clamoring for by 1995–96. To be sure, more than a few unsophisticated users had no idea that AOL *wasn't* the web. Everything "online" seemed the same to them. But other users began to forgo AOL's curated content for the freedom of the web. For AOL, this was troubling. The company had spent the better part of a decade and hundreds of millions of dollars building out its content offerings. Suddenly, it faced the prospect of users fleeing its online Eden.

AOL, CompuServe and Prodigy were all what the industry liked to call "walled gardens." They were online services that provided their users with proprietary tools and packaged content developed by the services themselves or their media partners. Little of what the online services did (with a few exceptions, such as email) interacted with the larger Internet, and none of the services were based on Internet standards. In a very real sense, online services like AOL didn't actually want users wandering outside of their networks and their control of the content. They much preferred if users stayed to play in the garden. The rise of the World Wide Web changed all this radically.

AOL always had a schizophrenic relationship with the Internet. The web provided a new, wilder alternative online environment, and in some ways this was in tension with AOL's carefully cultivated online "community." After all, would AOL prefer you researched cars on a *Car and Driver* "channel" on AOL proper, or by going on the web and visiting *Car and Driver*'s website? In interviews from the time, Case repeatedly floated the notion that the web was complicated and "niche" while AOL was targeting a mainstream audience by providing simplicity: "One disk to install. One price to pay. One customer service number to call. Building web sites and hoping people will find them is a significant leap of faith."[26]

"Their attitude toward the Web is a little grouchy," an industry researcher said of AOL at the time. "They have a hard time getting past their own resentment that this disorganized cousin is taking over in the public's mind. But it is a bias against inescapable realities."[27]

At the same time, however, the web also presented AOL with a rare opportunity. AOL's millions of users were still paying by the hour to dial in, and if AOL simply turned on access to the wider web, those same users would still be paying for the privilege of going through AOL's pipes. As AOL executive Ted Leonsis put it,

AOL could become "the Carnival Cruise Lines" of the Internet, the trusted guide to places unknown.[28] Prodigy was actually the first online service to allow its users to browse the web, in December 1994, but AOL soon followed suit.[29] AOL then rushed headlong into a $160 million Internet-based spending spree in order to keep abreast of the changing landscape.[30] A perfect example was BookLink and its Internet browser, which AOL snatched from Microsoft's clutches in November of 1994. AOL bought companies like Advanced Network & Services Inc. to build out its dial-up network (and thereby burnish its credentials as an ISP), and it bought a website called the Global Network Navigator, an early version of a search engine/Internet directory.[31] There were even very serious discussions about AOL doing some sort of investment in the young Netscape.

AOL's pivot to position itself as America's most popular on-ramp to the Internet quickly paid dividends. The number of subscribers grew to 6 million. Almost overnight, one out of every three people surfing on the Internet in the United States did so via AOL's dial-up lines.[32] This growth showed up on the bottom line. AOL recorded revenues of $1 billion for the first time in 1996, tripling what the business had brought in only a year before. AOL's stock had risen thirtyfold since its IPO; its market cap reached $5 billion.[33] While it still insisted on paying lip service to its own walled garden of content, AOL had wisely ridden the web's growth like a bucking bronco.

But the bronco was not always easy to ride.

Starting at 4 A.M. on August 7, 1996, AOL's services went down for nineteen hours.[34] The outage made front-page news around the country and made AOL the butt of jokes on late-night talk shows. For AOL, it was a major public relations black eye, but at the same time, a validation of how important the service had become in a few short years. This wasn't just an early adopter's playground anymore; AOL was how Americans were increasingly living their online lives every day. Imagine the chaos that would occur today if there were no email, no web, no anything online for nineteen straight hours. The Internet itself hadn't crashed, but America's ability to access it had. Suddenly, that was a big deal. The service outage came on the same day that NASA announced the discovery of indications of water on Mars, but AOL was the lead story on CNN.

Worse was to come. While AOL was now the country's largest Internet service provider, it was still competing in a crowded field. In addition to Prodigy, CompuServe and MSN, there were thousands of independent mom-and-pop ISPs spread around the country. An independent ISP didn't have the packaged content and proprietary chat rooms that AOL had. The indies gave users one thing: the Internet. You dialed in and you were on the web, quick and dirty. Increasingly, that seemed to be all people really wanted. To stand out from their online service brethren, ISPs competed on price. A low monthly fee of $19.95 got you unlimited hours of usage. This put quite a bit of pressure on AOL, which still depended on hourly rates and overages for the bulk of its revenue. Why was the Internet worth $2.95 an hour on AOL when you could browse unlimitedly elsewhere for a flat fee? The pressure from cheap competition threatened AOL's meteoric growth. In a quarterly report at the end of 1996, AOL announced signing up 2.1 million subscribers, but at the same time losing 1.3 million subscribers who fled the service for other ISPs.[35] AOL was still the market leader in terms of sheer numbers, but this competition and customer churn started to worry Wall Street.

The hourly fee structure was unsustainable. MSN announced in October of 1996 that it would provide unlimited access to its service for $19.95 a month, copying the business models of the independents. AOL had no choice but to follow suit. Starting with the December 1996 billing period, AOL announced that it would switch all of its users over to unlimited usage plans for the price of $19.95 a month. There were concerns internally about whether or not this move would kill AOL's hourly golden goose. "I had data and I had projections on how much money we would lose," Jan Brandt says. "We had so many people that were paying us 50, 60, $70 [a month]."[36] Flat-rate pricing would bring that to an end. More seriously, there were concerns about the network's ability to handle the increased usage that would inevitably occur. After all, members who had previously tried to limit their time to a few hours here and there could now, if they wanted, leave their America Online connections going 24/7. AOL testing suggested that actual usage would only increase 50% or so in an unlimited paradigm, and theoretically, the network could handle that.[37] But those assumptions were only taking into account existing users. Wasn't the point of flat-rate pricing to stop the churn, win back old customers, and maybe entice new ones? Steve Case told a *Wired* reporter that he thought the company had the infrastructure in place to handle "runaway growth."[38] He could not have been more wrong.

The very first day that user accounts were switched to "unlimited" pricing, member sessions leapt from 1.6 million hours to 2.5 million hours.[39] The numbers would only go up from there as, over the course of the month, more member plans were switched over. In addition, December was, of course, the height of the holiday season, and plenty of new computers were unwrapped as gifts that month. Now, with the promise of unlimited usage, all those bundled AOL trial discs were suddenly a lot more enticing. AOL signed up a record half-million members that December alone.[40] AOL's daily usage numbers were now up to 4.5 million hours each day.

There were too many people trying to log in all at once. The service couldn't handle it. Across the country, instead of the familiar guttural noises of the modem connecting, users began to hear only busy signals. Frustrated members would try over and over again to connect, hoping to get lucky. If users did get online, they tended to stay on as long as possible because there was no telling when they'd have the chance again. Once more, there was nationwide consumer outrage. The jokes began to circulate again about "America OnHold." CompuServe launched an advertising campaign to attempt to take advantage of its rival's misfortune, using the phone number 1-800-NOT-BUSY.

"We didn't really have an internalized grip on how important we were to people's daily lives," Jan Brandt says of the crisis. "What we didn't calibrate was the ferociousness of the response. It was crazy and it was really enlightening. It was like, 'Oh my God! People love us! They really love us!' Or, at that point, they love-hated us."[41]

In the end, AOL would spend hundreds of millions of dollars in a crash program that attempted to increase network capacity and bandwidth. Millions more were set aside to refund users and head off lawsuits and government scrutiny. Television ads were suspended so as not to encourage too many new sign-ups until the problems were fixed. Over the first few months of 1997, the busy signals slowly went away and service went back to normal. And the especially positive news was that the users stayed loyal. Even in the face of this well-publicized fiasco, user churn subsided. And, once they could actually use the service again, members did so in ever-increasing numbers.

AOL survived on the strength of its branding as America's online gateway. A lot of Americans didn't want any other way to get online; many didn't even know there was any other way. "Long lines are endemic at Disney World," a new AOL executive named Bob Pittman said. "Folks hate them. But offer Six Flags as an alternative and they look at you like you are crazy. They don't think anything is a substitute for Disney." AOL survived and continued to thrive for one reason, according to Pittman: "It's the brand, stupid."[42]

4

BIG MEDIA'S BIG WEB ADVENTURE

Pathfinder, *HotWired* and Ads

But what exactly was the big draw of the web? Why were people clamoring for AOL to add web access? Why had Netscape gotten a billion-dollar valuation and Microsoft revamped its entire corporate strategy? What exactly *were* people doing on the early web? Well, it was hard to say at the time, and maybe even harder to say twenty-five years later. The early web was sort of everything and nothing at the same time.

The entrepreneur and venture capitalist Chris Dixon has remarked that "the next big thing always starts out dismissed as a 'toy.'"[1] This is very often true with Internet technologies; a new site or a new tool can, on first encounter, seem gimmicky. *Why would I ever want to use/do that?* is many a first user's impression of the new. The web and the Internet itself engendered this reaction among many during its early days. At the time, the most enthusiastic net cheerleaders were touting it as a revolutionary medium that would completely change our lives. But there were still others who looked at the net and saw, yes, a toy. And we have to admit that these skeptics had a valid point of view, even with the benefit of hindsight. Because so much of the early web was decidedly amateur.

For example, one of the notorious early websites was the Netscape Fishcam, which was maintained by Lou Montulli, one of the original Mosaic six that Marc Andreessen and Jim Clark had recruited from the University of Illinois. This was, simply, a live webcam of a fish tank. Nothing more. It still functions to this day at Fishcam.com. It had a spiritual twin in the world-famous Trojan Room coffee cam, which showed a real-time image of a coffee pot in the Computer Laboratory of the University of Cambridge, England. It was a live video feed of a coffee pot. That's it. But for people on the early web, the fact that, at any time of day or night, you could see if a coffee pot, halfway around the world, needed to be refilled, that was just—kind of cool.

The list of early web ephemera could go on and on. There was a site that translated your name into Hawaiian; Cows Caught in the Web featured bovine trivia for no particular reason; Interactive Frog Dissection allowed you to dissect a virtual frog; Doctor Fun pioneered web comics; the Ultimate Band List (formerly the Web Wide World of Music)[2] listed information on bands and concerts and indie music in general;[3] the Frank Lloyd Wright Source Page tried to catalog pictures and analyses of every work the great architect ever created; Hiram's Inner Chamber provided info on Freemasonry; if you were a fan of the nineties cartoon *Animaniacs*, the Animaniacs Page! was obsessively complete; the Bonsai Home Page was all things tiny Japanese trees. The nature of the web made publishing so simple, anyone could publish a website about anything, and lots of people did.

But there were early websites with serious utility as well. The first commercial web publication was called *Global Network Navigator,* or *GNN*. It was launched all the way back in May of 1993, under the umbrella of the technology publishing company O'Reilly & Associates. O'Reilly published computer books and manuals, and in 1992 published *The Whole Internet User's Guide and Catalog*, one of the first books about the Internet targeted to mainstream users. One O'Reilly "associate," Dale Dougherty, was tasked with creating a rudimentary website to put the online-catalog portion of the book actually, you know, online. Dougherty continued to add layers of content until it functioned as a sort of online magazine, as well as a directory, listing cool websites in one of the first attempts to bring search and discovery functions to the web.[4] GNN would eventually be swooped up by AOL during its 1994–95 buying spree of early web properties. The Bureau of Labor Statistics maintained a website very early on to provide up-to-the-minute data on labor market trends. FedEx allowed customers to track the status of package shipments before most people even knew the web existed. Alamo Rent A Car was the first to allow users to book a car from its website. BankNet in Britain was the first bank to allow online account creation (if not actual online banking). Nature's Rose Floral Services allowed you to order flowers from the web. Classifieds began migrating to the web almost from the beginning, because for years there had been digital antecedents on the message boards and newsgroups of the early Internet and online services. The company that would eventually become Monster.com began life in 1994 as a site called The Monster Board.

Newspapers, for all their later reputation as being roadkill in the Internet Era, were actually prominent Internet pioneers as well. But then, publishers had more protodigital experience than almost anyone. For years, newspapers dreamed of electronic delivery of their product; digital would mean the elimination, or at least mitigation, of their greatest cost centers: paper, printing and physical delivery. Like the cable and telecom companies, they had sunk millions of dollars into digital experiments going back to the late 1970s. The dream of digital took its biggest step with Knight Ridder. The *San Jose Mercury News* was a Knight-Ridder publication, and it just so happened to be the hometown newspaper of Silicon Valley. Perhaps it was that proximity to the swelling tech revolution that led the *Mercury News* to launch Mercury Center in 1992. Mercury Center would offer the *Mercury News*'s regular content, but with more in-depth offerings online. It was sort of how shows or publications will now often say, "If you'd like to see the full interview, go online." Mercury Center's wares were designed to be complements to what the paper was already doing—content extensions. It carried press conference transcripts, wire stories that didn't make the printed edition, and legal documents and notices. Codes were printed at the bottom of stories so that readers could call or log in for the additional content. This cost $9.95 a month, and users without a computer could pay $2.95 a month for phone and fax service. In other words, you could have headlines read to you or faxed to your home or office. All of this was made possible via a partnership with America Online, which handled the monthly fees.

The Mercury Center was a small but genuine success. Newspapers around the country came to the *Mercury News* to see how the experiment was working out. In early 1994, the *New York Times* ran a profile on the Mercury Center noting that there had been 5,100 sign-ups, which represented a little under 20% of America Online's 30,000 subscribers in the Bay Area, albeit, less than 2% of the *Mercury News*'s 282,000 subscribers. The *New York Times* article noted that one key innovation was that reporters were urged to interact with readers about their stories. Bob Ingle, who led the Mercury Center initiative, told the *Times* reporter, "Our communication historically had been: 'We print it. You read it.' This changes everything."[5] It was a lesson that all media entrants to the Internet era would have to learn, or not learn, at their peril.

In the winter of 1994 when the Netscape Navigator browser came out, the Mercury Center quickly embraced the web. In January 1995, the *Mercury News* launched a website, with access originally $4.95 a month, though the paywall was later dropped in an effort to land more advertisers. The Mercury Center again found small but genuine success on the web, with thousands of new subscribers and $120,000 a month in revenue by its first year. By 1997, the website could claim 1.2 million monthly visitors. Under the Mercury Center's auspices, the *Mercury News* would continue to break ground, becoming the first daily to put the entire content of a given issue online while also being the first to use the site to break news, instead of waiting for the next day's edition. In April 1995 when the Oklahoma City bombing occurred, a photograph flashed across the wires that would become iconic, that picture you might remember of a firefighter holding a child in his arms. The Mercury Center immediately posted it to the website, over the objections of the photo editor, who wanted to save it for the next day's front page.[6]

Similar experiments were taking place in the magazine industry.

In 1995, the journalist and commentator Michael Kinsley launched a web-only publication for Microsoft (this was in the midst of Bill Gates's "hard-core" web obsession). Formerly an editor at *Harper's Magazine* and the *New Republic*, Kinsley very much intended the new publication, called *Slate,* to be a "magazine," complete with issues and publishing dates. In an early memo to staff, Kinsley wrote: "There should be a notional moment each week when we 'go to press' and 'hit the stands' (one and the same in this medium). I would say Friday midnight. This will allow us to summarize the week, and allow people to read us 'fresh' over the weekend."[7] Readers would be encouraged to print up articles and read them at their leisure. Within the editorial brain trust, there were actual debates over whether any normal person could be asked to read any piece over 700 words on a cathode ray screen without eye strain or boredom. At one point, Kinsley argued that each new piece or article would replace an older piece and the old piece would disappear forever. *Slate* launched with page numbers and a traditional table of contents, even though, obviously, numbering pages on the web was pointless. There were even debates about whether or not to allow hyperlinks in the articles, for fear of sending people away to other sites. Most of these callbacks to print media would be abandoned shortly after *Slate* launched.

One of the most prominent pioneers of professional online content was the 800-pound gorilla of the media landscape: Time Warner. Rising from the ashes of the information superhighway fad, the seeds of Time Warner's groundbreaking web efforts came from the failed Full Service Network project in Orlando, Florida. Some of the same personnel involved in FSN would attempt to succeed on the web where interactive TV had failed in the home. The project would be led by Time Warner's Jim Kinsella, who would later go on to help develop MSNBC, and overseen by Walter Isaacson, now most famous for being Steve Jobs's biographer, but later also the editor of *Time.*

Time Inc.'s Pathfinder.com launched on October 24, 1994.[8] The site logged a reported 200,000 hits in its first week (back then, they called them hits) and reached 3.2 million pageviews weekly in its first year.[9] Pathfinder would have the requisite bulletin boards and chat, based on a proprietary system called WABBIT, and experiments also took place in the area of commerce, via a partnership with the ecommerce pioneer Open Market. But from the very beginning, the site was designed primarily to be a vehicle for showcasing existing Time Warner media content. Time Warner had experimented in recent years with licensing especially its magazine content to the likes of AOL and CompuServe, so, in a sense, Pathfinder was an attempt to do an end-run around the online services.

But the web was a different sort of animal, as Time Warner would learn to its great consternation. Time Warner wasn't quite comfortable with people talking back. Early on in the site's history, the O. J. Simpson case was raging and Pathfinder found huge success with an O. J. Central section where users could debate the case. But Time Warner executives feared that the free-for-all of comments and user-generated debates might expose the company liability-wise. Management slowly began to discourage community efforts. Community editors were tasked with policing community sections, discouraging users who saw their comments censored or deleted.

Attempting to cram its entire portfolio of brands into one awkwardly constructed website also proved unwieldy. Time Warner had a stable of world-class content, but instead of leveraging those brands (some of them, like *Time*, *Sports Illustrated* and CNN, among the most trusted media sources in the world), everything was cobbled together under Pathfinder's big tent. You couldn't get *People* magazine content by going to People.com. You had to go to Pathfinder.com/people. And some of the executives at these brands resented that state of affairs. For years, *People* magazine refused to mention its web presence in print, losing Pathfinder millions of dollars in free publicity. Because the various brands resented the Pathfinder umbrella, and because Time Warner had such a vicious culture of warring fiefdoms and corporate politics and infighting, the various brands tended to withhold their best content from Pathfinder. Some of the sharpest fights among management centered on how prominently to make the online masthead. "The situation really did remind me of Italian city-states," Bill Lessard, a Pathfinder producer, said. "A loose confederation warring against each other and against [Pathfinder]."[10]

In the end, despite being the most aggressive and well-funded early content pioneer on the web, Time Warner simply couldn't make heads or tails of how the web could work as a publishing business. "We're all looking at the elephant," said Time Inc. editor in chief Norman Pearlstine at the time, "but what people think we should do depends on what part of the elephant they're looking at."[11] Simply by virtue of being a pioneer, and almost in spite of its own dysfunction, Pathfinder achieved a considerable degree of success in terms of audience.[12] But it was also a huge money loser just as the Full Service Network had been. Time Inc. chairman and CEO Don Logan gave Pathfinder an infamous preemptive epitaph when a reporter asked him about the site's financial performance. Pathfinder had "given new meaning to me of the scientific term 'black hole,'" he quipped.[13] Pathfinder suffered a slow, ignominious death march before being officially shuttered in 1999. Estimates for the cost to Time Warner over the course of Pathfinder's life range from $100 million to $200 million.

Mercury Center, *Slate*, Pathfinder and the others were part of a larger process of journalists and professional media folk finding what worked and what didn't in this new medium. On the web, you could publish something when you had something to publish, and it was instantly interactive. But the greatest lesson to learn was how to make this unruly web pay. The solution to that problem would come from another magazine dabbling in the World Wide Web, and in the process, the very business model of the larger Internet would be discovered: advertiser-supported content.

■

PEOPLE LIKE TO THINK of *Wired* magazine as being the harbinger of the Internet Era; but, in fact, *Wired* predated the web going mainstream. The brainchild of Louis Rossetto and his partner, Jane Metcalfe, *Wired* revolutionized print media by embracing, as its subject, and in its very design ethos, a promised digital future of limitless possibility and technological utopianism. *Rolling Stone* for the computer era, the magazine peddled the "radical Libertarian" Rossetto's vision of a digital revolution that would set mankind free in both physical and spiritual ways.

Wired didn't foresee that the digital revolution it was hawking would take the form of the World Wide Web any more than Bill Gates did. Once the web began happening, however, *Wired* quickly became a vocal cheerleader and tried to embrace the new platform in deed, not just in rhetoric. In early 1994, *Wired* hired a young financial wizard named Andrew Anker from the investment banking firm Sterling Payot, which had been instrumental in putting together the magazine's initial funding and financing. Hiring a numbers guy made sense because Anker's remit from Rossetto was to make sure whatever online experiments *Wired* dabbled in would pay for themselves.

"My mandate was: We're building a business here," Anker remembers.[14] Anker wrote a business plan and launched the new enterprise under the rubric Wired Ventures, a separate company within the Wired umbrella with Anker himself as the CEO. Anker led the development of a website called HotWired.com, which would be some mix of existing magazine content, along with original reporting and multimedia features that would attempt to take full advantage of the web's interactive nature. There was a brief flirtation with the idea of a paywall, or limiting the site to existing subscribers. But *Wired* was a magazine flush with success primarily as an advertiser-supported operation, so, the *HotWired* brain trust logically turned to the notion of merely replicating the model they already knew. It was decided that *HotWired*'s launch on the web would be sponsored by advertisers *Wired* had existing relationships with from the print side of its operation. Like the launch of a print magazine, advertisers would be asked to sponsor specific content sections of the new website for a flat fee. "$10,000 was a round number that made the numbers work," Anker remembers. "And we tried it and everybody sort of seemed to buy it."[15]

If it sounds like they were feeling around in the dark, it's because they were. Nothing like this had been attempted before. The first genuine advertisement on the World Wide Web was published by Global Network Navigator, which, in 1993, sold an ad to a Silicon Valley law firm, Heller, Ehrman, White & McAuliffe. It was text only, a glorified classified listing. Later, GNN sold the first sponsored hyperlink, pointing to a children's catalog retailer called Hand in Hand. Clicking sent a user to the company's rudimentary web page to learn more about Hand in Hand's strollers and cribs.[16]

But those experiments were simply one-off, cash-for-placement deals. The *HotWired* team was attempting something more ambitious, both technically and aesthetically. Two advertising and digital design firms, Modem Media and Organic, were brought on board and tasked with designing and selling something that felt closer to a magazine-style ad. Big. Colorful. Eye-catching. These would be the very first banner ads.

Joe McCambley was a creative executive at Modem Media. "I remember having a big debate—and we probably argued for an hour or so—about whether or not it should even be a color ad," McCambley says. "We knew we could make it smaller [in terms of bytes] if it were black and white. We knew there was a large percentage of people out there that only had black and white monitors anyway."[17]

"At that time, you couldn't actually even center a banner," remembers Organic's Jonathan Nelson. "Everything was flush left. You would make the banners only two or three different colors. And you couldn't have complex graphics in them because everybody was on modems at the time. Bandwidth was extremely limited."[18] If a graphical ad took two minutes to download onscreen, no one would read the article, much less see the banner ad.

"The size of the ad was really created because of the size of the browser at the time, and the scroll bars on the side, and people just trying to figure out exactly what would fit," remembers Craig Kanarick, then a multimedia designer who worked on the first ads. "Somewhere around 460×60 was the right number [in pixels]."[19]

But there was an even deeper philosophical problem to solve. With magazine, radio, television, even billboard advertising, the ad merely made an "impression" on the audience. It was a passive thing. The web was decidedly not passive; the web was about links, about clicking. So, what should these new banner ads do? What would happen if users clicked on them?

"Not only were there not a lot of really big corporate websites, at the time there was really a debate whether a corporate website would actually be the thing that people wanted," says Kanarick. "Like, who would want to go to a website like Pampers, where they're just going to talk about diapers all the time?"[20]

HotWired launched on October 27, 1994, before Netscape's first beta browser, before Pathfinder, before *Slate*. And it launched with a full roster of banner ads from the likes of AT&T, Sprint, Timex, MCI, Volvo, a modem company called Zircom, as well as that infamous carbonated alcoholic beverage of the 1990s, Zima. The banner ads ran in skinny rectangles above, below and within the content. There's really no singular "first" banner ad, because all the sponsorships launched simultaneously. In retrospect, however, the advertising industry likes to think of the AT&T ad as the default "first" because the copy of the ad was certainly prescient. It read: "Have you ever clicked your mouse right here? YOU WILL."

This was part of the family of nationwide television and radio ads that AT&T was blanketing the country with that year. The TV ads, narrated by Tom Selleck and directed by David Fincher, all had the convention of "Have you ever done x?" followed by the assurance that "You will. And the company that will bring it to you is AT&T." So, for example, one ad had a mother tucking her children in via video-call ("Have you ever tucked your baby in from a phone booth? You will . . .") and another showed in-car GPS navigation, pretty much as it exists today ("Have you ever crossed the country without stopping for directions? You will . . .").

"They showed this sort of Jetsons future of the world," says Craig Kanarick. "Many, if not all of those, have come true. But at the time it was really this sort of fantasy about how the future is going to be amazing."[21]

The first banner ads got click-through rates in astronomical percentages. "People just clicked on anything to see what might lead them somewhere," Joe McCambley says. "It bordered between the high 70s and low 80s* [in terms of click-through percentages] for about 2–3 weeks."[22] Andrew Anker concurs: "People were clicking on every single page. And ads were just as interesting content as our content."[23] In no time, the other media sites—Pathfinder, *Slate*, etc.—were following *HotWired*'s lead. Pathfinder would, in fact, launch with only one advertiser, AT&T, running some of the same "you will . . ." ads.

Functionally, the first banner ads were an introduction to the way the future was going to work, at least on the web. To this day, most of what we do online, with the exception of ecommerce and the rare subscription service, is all advertising-supported. It's conceptually jarring to realize that a medium and an industry that we think of as being so futuristic and technological is sustained by a business model that is centuries old. But then, one of the very first things that the web disrupted was advertising itself, because the Internet and the web promised to revolutionize advertising in ways that marketers had only dreamed of previously.

■

IT HAS ALWAYS BEEN devilishly hard to measure the actual effectiveness of advertising. John Wanamaker, the department store mogul, famously said, "Half the money I spend on advertising is wasted; the trouble is I don't know which half." You can pay for an ad in a magazine, but you'll never know how many readers actually flip to that page and see the ad. And if a reader does flip to the page, how do you know if he or she actually reads the ad? The same is true for advertising on radio, television, movies, even billboards. An advertiser can buy a billboard on a highway that carries 30,000 commuters every single day. But who knows how many drivers look up and take notice? This is why advertisers have always been obsessed with things like circulation numbers and ratings points. An ad is only effective with a small percentage of an audience in the first place, so the best way to spend money effectively is to try to reach the largest audience of likely customers.

---

* In the modern web era, a click-through rate of 0.5% is considered a blockbuster.

Online advertising promised to make this vague science obsolete. Because a computer serves up webpages, on the web it is possible to know the exact number of times a web page—and with it, a given advertisement—is delivered to an audience. No more guesswork. An advertiser can know to the second, and often in real time, when their given block of 1,000 ads has been served up. Furthermore, the web allows an advertiser a better gauge of how many people ignore a given ad. Because each ad is clickable, often leading to the advertiser's own website or another traceable property, the web allows an advertiser to measure how many people interact with an advertisement. They can know how much of an impression the advertisement makes on an audience. In the language of the advertising business, this is called "engagement."

Beyond even this, there are the cookies, those little lines of software code that follow you around the Internet once you visit a website or click on a web ad. Cookies were first developed by Netscape's Lou Montulli and included in the first versions of the Navigator browser. Cookies were originally intended to add "memory" to the web, allowing users to remain logged in to sites and to refresh content so that they wouldn't be served the same thing every time they returned.[24] But publishers like *HotWired* latched on to this technology as a way to deliver targeted ads to specific audiences. What users have been frequenting windsurfing websites recently? Cookies can tell you, as well as any advertisers who want to market to windsurfing enthusiasts. Add to this the voluntary information audiences online might be willing to share with a given website. Your name, your age, sex, income, geographic location—in ways complex and yet not entirely appreciated, our online activities have delivered the holy grail of advertising from time immemorial: knowing exactly an audience's interests so that the ad man can market only to the most promising leads.

The web seemed like the advertiser's promised land. Each time a webpage was loaded, this would be counted as an "impression," for which an advertiser would pay on a CPM basis.* But what the

---

* For decades, all advertising has been sold using a metric called CPM, or cost per mille. In Latin, *mille* means thousand, so CPM is essentially saying that an adver-

advertiser was really after were the "clicks." Measuring the "click-through rate" provided a greater measurement of advertising engagement. Forget mere passive impressions, it was now possible to measure how often a user *interacted* with an ad. In the age of ecommerce, advertisers could even measure clicks that led directly to a sale. This made it easier for advertisers to calculate their return on investment by orders of magnitude. For the first time it was possible to know which half of the advertising spend was wasted.

Another advantage of the web was how it fit into the historical advertising paradigm. There are only so many hours in a day. So, broadly speaking, advertisers are interested in how many hours of the day a given medium can capture a person's attention. How many hours a day does the average person listen to the radio? Read the newspaper? Watch television? Advertisers—especially the larger ones—apportion their overall advertising "spend" based on what percentage of a person's daily attention they can capture. The Internet represented the first new advertising medium to come along since the advent of television. As Americans came online in increasing numbers, the Internet promised to capture more and more of their time and attention. It was expected that advertisers would logically shift their advertising spend to try to advertise against this new attention center. And sure enough, the ad money chased the eyeballs. In 1995, around $50 million was spent on banner advertising on the web.[25] By 1997, online advertising passed $1 billion for the first time.[26] That was a mere rounding error compared to the $60 billion corporations spent on advertising across all mediums that year. But everyone anticipated that online ads would grab an ever-growing slice of that lucrative advertising pie.

They were right.

In 2015, digital advertising hit $59.6 billion.[27]

---

tisement is priced based upon how many thousands of people are exposed to a given advertisement. Imagine that the total cost of running a full-page ad in a magazine is $50,000. Now, imagine that the magazine has a circulation of 4 million people. $50,000 divided by 4 million is .0125. CPM is calculated by multiplying .0125 by 1,000. So in this example, the advertisement in question has a CPM of $12.50. The advertiser is paying $12.50 to reach every thousand readers of that magazine.

5

HELLO, WORLD

The Early Search Engines and Yahoo

Soon after *HotWired*'s launch at the end of 1994, it was estimated that the number of websites in the world had passed 10,000.[1] But even though "professional" sites like *HotWired* and Pathfinder were beginning to proliferate, the vast number of websites and webpages remained random, even individual, affairs. Most of the early websites had to publish wherever they could, and that often meant piggybacking on existing academic or corporate websites. It wasn't until 1995 that individuals were broadly allowed to register their own .com domain names.[2] So, if you wanted to visit Apple's website, you could go to www.apple.com. But if, say, you were looking for Gabriel's HTML Editor List to find good HTML-authoring software, you had to browse to http://luff.latrobe.edu .au/~medgjw/editors/. If you wanted an online tarot card reading, you had to type in http://cad.ucla/edu/repository/useful/tarot.html.[3] This inscrutability combined with the web's vastness and anonymity presented a tree-falling-in-the-woods sort of problem. Anyone could now publish anything; but if you did, how would anyone ever know about it?

And so, necessity dictated that search engines would become the most popular and most important early websites. And because the problem of a business model had been solved by *HotWired*, search sites, and Yahoo in particular, would become the web's first great companies.

There were many different early web search engines and tools, and they all had varying degrees of utility.* The not so secret truth about all the early search engines was that they weren't very good. They returned results in a way that could be comprehensive, but often had no accuracy. A search for, say, "windsurfing" might give you a list of every webpage in the world that mentioned the word "windsurfing," but made no effort to sort for context. What was the best windsurfing site on the web? The search engines had no way of telling you. A more refined search for, say, "windsurfing in California" might return sites for windsurfing *or* California, but maybe not both. The searcher might find the State of California's official government site at the top of the list, or a site for a windsurfing company in Hawaii.

The cause of these poor results came down to the automated nature of the search process itself. To this day, a "search engine" is actually a database of website copies. The search engine sends out "spiders," which are computer programs that go out onto the web and find new web pages. The spiders locate the pages and then copy some or all of the code into the search engine's own database. When a user searches a search engine, they're not actually searching the web itself, but are instead querying the database of copied webpages the search engine has compiled. The accuracy and comprehensiveness of this database varied from search engine to search engine and the results therefore varied depending on the weight each search engine gave to various factors in the database. Search engine A might list a certain webpage as the number-one result for windsurfing because the word "windsurfing" was prominent in the title of the webpage. But search engine B might list a completely different page as the first result because the word "windsurfing" showed up the greatest number of times in the body of the page.

---

* Pinning down which one was first is open to debate. For the sake of brevity and clarity we can focus on those that were the longest-lasting and actually led to websites that would become familiar to everyday web surfers.

Though they worked hard to make them otherwise, the algorithms the early search engines used to sort and rank pages were crude and wildly ineffective. The obvious alternative to this state of affairs was to bring a curatorial element to search. And in fact, the dominant player that would emerge in search was not strictly a search engine at all, but a directory, compiled not by bots but by actual humans.

In early 1994, Jerry Yang and David Filo were Ph.D. students in electrical engineering at Stanford. They knew each other from their studies but really bonded when they signed up for a brief teaching stint in Japan. The dissertations that the two were (ostensibly) working on in the spring of 1994 involved design automation software, which was a hot area of research at the time. Yang and Filo shared side-by-side cubicles in a Stanford portable trailer, in lieu of official offices. Their dissertation advisor was on sabbatical, so they were free to order pizza, goof around, and, oh yeah, occasionally research. More often than not, one or both of them would end up sleeping in the trailer. A friend called the trailer "a cockroach's picture of Christmas."[4]

The two students weren't exactly burning through their dissertation. Filo had discovered the Mosaic browser shortly after it was released, and this led to an all-consuming obsession with the World Wide Web. In those days, it was still possible to visit every single website in existence in a matter of a few hours. But new websites were popping up every day. Always a bit competitive, the two began collecting and trading links to the new websites they found. They started compiling these favorite links into a list, each trying to outdo the other by finding the coolest new site of the day.

This was right at the moment when Mosaic was lighting the fuse under the powder keg that was the web. As the web grew that summer, things got a bit more complicated. Because Yang's workstation was hooked up to Stanford's public Internet connection, other people could view the list the two were generating by going to http://akebono.stanford.edu. The list was called "Jerry's Guide to the World Wide Web," and it proved popular among Yang and Filo's group of friends. Word of mouth spread news of the list even further, and soon complete strangers were emailing suggested websites for inclusion. In order to keep things reasonably organized, Yang and Filo broke the list out into a hierarchical directory. Thus, to find MTV's home page, a user drilled down by category: Entertainment > Music > Music Videos > MTV.com. The pair came up with their own software to seek out ever-newer sites and webpages, but the additions to the directory were made entirely at Yang and Filo's discretion. In those days, there was no automation or algorithm.

The pair began working on the directory to the exclusion of almost everything else. They would toil away for dozens of hours at a stretch, trading off sleeping on the floor. For Yang and Filo, it wasn't work; it was fun. "We wanted to avoid doing our dissertations," Yang admitted.[5] By September 1994, Yang and Filo had compiled a directory of more than 2,000 sites. What was more impressive was the fact that Jerry's Guide to the World Wide Web was getting 50,000 hits (searches) a day. "We were in a unique situation in the summer of 1994," Yang remembered later, "to be able to experience that kind of grass-roots growth, fueled by a lot of interest that was not our doing, and then just sitting back to watch the access logs go up."[6] The pair decided that their project needed a better name. A convention among software developers at the time was to name projects "Yet Another Something Something." For example, YAML was Yet Another Markup Language. So, Yang and Filo settled on the name Yahoo!, which they claimed stood for Yet Another Hierarchical, Officious Oracle. The exclamation point was irreverent and entirely intentional—as Filo put it, "Pure marketing hype."[7] The URL became http://akebono.stanford.edu/yahoo.

Stanford has a long history of supporting student-run projects that may or may not evolve into startups. So, at least initially, the university was a generous host of Yahoo's traffic and content, free of charge. When Netscape launched its beta browser late in 1994, it decided to make Yahoo the default link when a user clicked the DIRECTORY button on the top menu of the browser. No one could have anticipated it beforehand, but having a button in Navigator's menu bar was almost as valuable as having an icon on the Windows desktop. The flow of curious web searchers grew into a flood. Yahoo had its first million-hit day late in 1994. By January 1995, Yahoo had grown into a directory of 10,000 sites and was getting more than 100,000 unique visitors a day. The servers began to struggle under the deluge, so the university asked Yang and Filo to find another host for their website.

For Yang and Filo, it was the moment of truth. For months they had left their dissertations languishing. Now it was time to decide if Yahoo was a real thing or not, and whether or not the "boys" were willing to become businessmen. "David had it in his gut very early on that Yahoo could ultimately be a consumer interface to the Web rather than simply a search engine or a piece of technology," Yang told *Fortune*. "We weren't really sure you could make a business out of it though."[8] Interested parties were already forming a line at Yahoo's trailer door. Reuters, MCI, Microsoft, CNET and a pre-IPO Netscape all met with the creators to see if some form of partnership or buyout was possible. "I remember sitting in their trailer in December of '94," remembers Tim Brady, who was a friend of Jerry and David and would be one of their first hires. "And they had a voicemail system, and the head of the *Los Angeles Times* was calling, AOL was calling, and those were just the ones that were on the voicemail that day."[9]

The venture capitalists soon came calling as well, and now that they needed a permanent home, the boys were ready to talk. But the moneymen were skeptical about Yahoo's chances of being a sustainable business. Netscape might have seemed like a dubious proposition when it was looking to raise funds: barely making money, kinda-sorta giving away its product for free, unproven market, etc. But at least Navigator was a software package. People understood that software could be sold. Netscape was proving it could make real money providing support and server packages to supplement its software. Yahoo, on the other hand, was a service; a destination; a directory; a glorified list. There was almost nothing proprietary about it. Furthermore, it was a service that you could never charge for. Yang and Filo were convinced—quite rightly—that the day they started charging users to search would be the last day users ever visited Yahoo again. If Netscape's business seemed intangible,

Yahoo's seemed downright hypothetical. Yang began circulating a scratched-together business plan, but this failed to impress the VC's who were sniffing around.

One of those who made the trek to the messy trailer before Yang and Filo vacated Stanford was a VC named Mike Moritz. Moritz described the squalor as "every mother's idea of the bedroom that she wished her sons never had."[10] He and his team quizzed Yang and Filo among the empty pizza boxes and humming workstations, asking the obvious question: "So, how much are you going to charge subscribers?"[11]

"Dave and I looked at each other and said, 'Well, it's going to be a long conversation,'" Yang would recall. "But two hours later, we convinced them that Yahoo should be free."[12]

Moritz was a general partner at the VC firm Sequoia Capital. Sequoia had funded such Silicon Valley luminaries as Apple, Atari, Cisco and Oracle, but it had not yet dipped its toe into Internet waters. The vision that Moritz used to argue Yahoo's case was the one put to him by Yang and Filo. It sounded like a mix of the Netscape strategy with a bit of AOL sprinkled in. Yahoo already had millions of loyal users; surely there would be some way to monetize them. As more and more users were coming to the web, Yahoo could be the friendly guide that would hold the hands of new users and lead them out into the void. If there was an elevator pitch, it was that Yahoo had the chance to be the *TV Guide* for the Internet. Like Yahoo, *TV Guide* simply provided information that any other entity could aggregate. And yet, *TV Guide* was (at that time) the largest-circulation magazine on the planet.

Sequoia eventually bought the pitch. Yahoo already had an audience of millions, and if the web kept growing at its present rate, who knew how many hundreds of millions could be reached in the near future? By that logic, even the wacky company name could be seen as a plus. After all, as Don Valentine, the legendary founder of Sequoia, put it, "A long time ago, we helped finance a company called Apple."[13] Sometimes investments in companies with silly names could turn out handsomely.

In April 1995, Sequoia invested $1 million in exchange for one-fourth of the newly incorporated Yahoo. By early 1999, Sequoia's initial $1 million was worth $8 *billion*.[14]

With their first infusion of cash, Yang and Filo secured 1,500 square feet of office space at the auspicious address of 110 Pioneer Way.[15] Engineers were brought onboard to help Filo set up Yahoo's servers and technologies in-house. The Yahoo.com domain was registered. Finance folk were brought on to structure Yahoo like a lean, mean startup. An "adult" was brought in to be CEO, in the person of Tim Koogle, a veteran of both tech startups and the tech establishment, in the form of Motorola. As for the two founders, Yang took the official title of "Chief Yahoo" and continued to be the face of the company. Filo took the title "Cheap Yahoo" and dedicated himself to keeping the tech side running smoothly and frugally. Most important, a cadre of new hires was fashioned into a team of professional web surfers who would help build out the Yahoo directory and stay on top of the exploding web. The surfers, who would eventually number more than fifty, were each expected to add as many as a thousand new sites a day to the directory.[16]

The web was growing exponentially, and Yahoo needed to keep up with it. But it also had to keep looking back over its shoulder. If a deep-pocketed competitor copied Yahoo's glorified list, what could prevent Yahoo from being steamrolled? "It wasn't rocket science," Filo admitted. "We didn't have patents or anything like that. Someone smart with resources could have done the same thing."[17]

For his part, Jerry Yang was confident that Yahoo had one unique advantage: it had been first. It would become an article of faith during the dot-com era that being early to market on the Internet frontier conferred a magical "first-mover advantage" on whomever was so fortunate. Certainly, Yahoo's experience did nothing to disprove this. Those early months as the default search tool on Navigator had sown the seeds of familiarity and loyalty among early Internet adopters. Even when competing services showed up, users had a tendency to stick with what they knew, so long as it continued to work. The first-mover advantage meant that Yahoo had a big head start in the land grab for market and mind share among early web devotees. This was a lead that was Yahoo's to lose. In order to stay ahead, Yahoo decided to take a page out of AOL's book. It would brand itself in order to reinforce its users' loyalty.

With millions of users already familiar with Yahoo and tens of millions of "newbies" on their way, becoming the first Internet brand would be invaluable. Karen Edwards was brought on to direct Yahoo's marketing efforts. With previous experience at Clorox and 20th Century Fox, Edwards bought an offline-industry faith in the power of branding to the new world of clicks and browsing. From her very first interview with the company, Edwards pushed the idea that building a strong brand might create a defensible moat around Yahoo's unpatentable and eminently copyable service. "I think we could really make Yahoo a household name," Edwards told her new coworkers. "I remember Jerry Yang laughing, 'Ha, ha! A household name?' "[18] But under Edwards's direction, Yahoo did something that was completely radical for the time: advertise on TV and radio. Yahoo was the first Internet company to market itself via mass media. With zippy, hip ads, matching the slick name and the brash image of the site overall, Americans found themselves being asked "Do You Yahoo?" Yahoo quickly became one of the Internet's most recognizable names, familiar even to the vast uninitiated Americans who were not yet even online. With its quirky purple logo, Yahoo was soon everywhere, from hockey rinks to billboards to t-shirts. *Businessweek* said that Yahoo projected a "cool California image—hip but not rad, easy-to-use but not simplistic."[19] In the twelve months after starting the "Do You?" campaign, traffic to Yahoo's website quadrupled.[20] By 1998, Yahoo was better known to the average consumer than even Microsoft.

"The fundamental bet we are making is that we are a media company, not a tools company," Yang told *Fortune* magazine. "If we are a tools company, we are not going to survive. Microsoft will just take over our space. If we are a publication, like a *Fortune* or a *Time*, and we create brand loyalty, then we have a sustainable business."[21] Making Yahoo the first great brand of the Internet Era would serve the company well throughout the entire dot-com period. When later asked why Yahoo enjoyed a greater stock market valuation than rival search companies such as Excite, a stock analyst would reply, "Yahoo is cool! It's not a technology company. It's a brand, it's an article of culture."[22]

Then came the Netscape IPO in September 1995. The Internet was hot and Wall Street was in search of other net companies that seemed to have the same growth trajectory. Search engines had the largest audience of netheads anywhere, and Yahoo was leading the pack. By February of the following year, the site was seeing more than 6 million visitors every single day.[23] Those traffic numbers were double what Yahoo had seen just five months before. The growth was parabolic.

Now that Wall Street was living in a post-Netscape world, the pressure built for Yahoo to go public as well. They didn't need to; additional rounds of investment left Yahoo with quite a war chest. But Yahoo's competitors, the search "engines" Excite, Lycos and Infoseek, were all filing to go public in Netscape's slipstream. Yahoo couldn't turn down the opportunity to raise even more money and maintain its lead against its search rivals. Plus, Netscape had shown that there was an incredible amount of free publicity to be gained by a successful, high-profile IPO. By sitting out the party, Yahoo risked ceding its role as the industry leader, at least in the eyes of Wall Street.

Excite and Lycos enjoyed moderately successful IPOs in early April 1996 (Infoseek went public a few months later). Yahoo went public on April 12, selling 2.6 million shares, initially priced at $13, but seeing a first trade price of $24.50.[24] Over the course of the first day, the stock peaked at $43 before ending the day at $33.[25] This 154% leap over the offer price was better than even Netscape's 105% first-day pop.[26] More important, this made Yahoo's market value $850 million, which was more than Excite's $206 million and Lycos's $241 million combined.[27] As planned, Yahoo's IPO made all the other search sites look like pretenders to the throne.

Yahoo now had hundreds of millions of dollars in the bank. Yang and Filo had each pocketed about $130 million on paper. But Yang said that the IPO had merely induced "panic—no, not panic, but anxiety."[28] That was because there was one looming problem: for all the dollar signs in Yahoo's bank account, there wasn't actually much on Yahoo's bottom line. In its first quarter as a public company, Netscape had recorded revenue of $56.1 million.[29] By comparison, in *its* first quarter as a public company, Yahoo could report revenue of only $3.2 million.[30] Even that was better than 1995, when Yahoo reported revenue of only $1.4 million *for the entire year.*[31] Again, if Netscape had gone public with questionable revenues, Yahoo had taken things to the next (more speculative?) level. But investors had shown that they were willing to invest in unprofitable young web companies as long as they could show growth. Yahoo would be okay as long as it could show continued audience growth *and* as long as it could find a way to monetize that audience. One day. Preferably soon.

In the meantime, of course, *HotWired* had shown the way to easy money: web content could be subsidized by ads. Unlike *HotWired* or Pathfinder or *Slate*, in Yahoo's case, it didn't even have to produce the "content" on its site itself. The content was the web! Yang and Filo didn't want ads to interrupt their directory, but ads *around* the directory, sort of like the ads around the *HotWired* articles, might be okay. At the time, Yahoo liked to give the impression that it came to the advertising model reluctantly, but really, there was no other feasible option available to the company. As early as April 1995, soon after the original Sequoia investment, David Filo granted an interview to *Advertising Age* magazine. Under the headline "A Gaggle of Web Guides Vies for Ads; Yahoo Directory Opens to Sponsorship Deals as Competition Grows," Filo declared, "Because we are now backed by a third party, there's pressure to produce. Yahoo will have to become a money-making enterprise. We're not sure if we want to start reviewing sites or continue to just list sites in a comprehensive fashion, but we are definitely going to integrate advertising into what we do."[32]

Yahoo treaded lightly, putting a survey on its home page asking users whether they would countenance ads. The response was lukewarm acceptance. Nevertheless, there were those inside the company who feared that even introducing graphics might fundamentally alter the freewheeling ethos that made Yahoo unique. When the first ads were launched later that month, according to

Tim Brady, "The email box was immediately flooded with people badmouthing us and telling us to take it off. 'What are you doing? You're ruining the net!' "[33] The Yahoos held their breath to see if the ads chased searchers away. But the protests quieted down after only a few weeks. The directory was just as helpful as it always was. The users stayed loyal.

Once Yahoo turned on the advertising spigot, it ramped things up rapidly, signing on more than 80 sponsors in less than six months.[34] The advertisers and the advertisements would only increase with Yahoo's growing traffic numbers. By 1996's fourth quarter, the website could boast 550 advertisers, including many Fortune 500 companies such as Wal-Mart and Coca-Cola. This all led to an impressive 1,300% increase in its revenues, to $19.7 million in 1996. But because the web was growing every day, the company found it literally could not sell ads fast enough. By the end of 1996, as pageviews reached 14 million a day, as much as 75% of Yahoo's potential ad space went unsold.[35] There was simply too much traffic to sell.

Because Yahoo had so successfully branded itself as the Internet's version of the Yellow Pages, countless brands and retailers jockeyed to purchase valuable real estate on Yahoo's directory. New dot-com companies would compete viciously among themselves for prominent placement. Amazon.com and CDNow.com could be played off one another to advertise music sales alongside Yahoo's Music categories. E*Trade and Datek online would sign multimillion-dollar deals just to put online trading buttons in Yahoo's Finance sections. And it wasn't just retailers: when Yahoo decided to add news, weather, stock prices and other curios to its directory, it found that media partners such as Reuters were eager to partner and provide content in exchange for a share of the advertising revenues.

"There was a land grab," a Yahoo marketing executive would remember. Yahoo was perfectly positioned to take advantage as Internet mania took off. "It was no one's fault, but lots of companies were overinvesting and trying to grow too fast. It's hard to blame Yahoo for that—but sure, we were right there taking the money."[36] By 1997, the online advertising market neared $1 billion, and Yahoo alone was estimated to control 7.5% of the total.[37] Yahoo's advertising base shot to 1,700 brand clients. These advertisers were chasing traffic that had skyrocketed to an astounding 65 million pageviews per day. And all of this led to a proportionate 257% rise in revenues to $70.4 million.[38] Yahoo's stock rose accordingly, jumping 511% over the course of 1997. The company at that point had a market value of almost $4 billion.

Yahoo was bigger than Netscape. But unlike Netscape, which remained a traditional software and business services company, Yahoo was a web-only company, a web-native company, a company that would never have existed if the web had never been invented.

6

GET BIG FAST

Amazon.com and the Birth of Ecommerce

If the code had finally been cracked in terms of making money on the Internet, then it seems inevitable that people would eventually use the web to sell things. There was more than a century of precedent for doing commerce remotely: the multibillion-dollar catalog sales industry. A webpage could be a more dynamic and effective catalog than what Sears or Lands' End could offer. And the Secure Sockets Layer technology developed by Netscape made actual transactions possible on the web; no need for 1-800 numbers or customer service reps to take the orders. Indeed, perhaps the longest-lasting legacy of Netscape Navigator setting the standard for the early web is that, to this day, SSL, via its descendant, TLS, enables the vast majority of online commercial transactions worldwide.

But if you were willing to look deeper at the opportunity presented by web commerce, then you could envision even greater possibilities, even greater efficiencies and economies of scale. Just as the newspaper industry dreamed of delivering its product without the need for costly delivery and production expenses, a forward-thinking retailer could dream of a world without the need for costly commercial real estate expenses and perhaps vastly simplified warehousing and logistics costs. To early commerce pioneers, the promise wasn't that the web would allow them to do something fundamentally different than before—this was still about selling goods to consumers—but that it could radically transform the *way* they would do it.

Nearly twenty-five years on, this vision has largely come to pass, and in the popular imagination it has come to pass because of one company. Pioneers of new technologies are rarely the ones who survive long enough to dominate their categories; often it is the copycat or follow-on names that are still with us to this day: Google, not AltaVista, in search; Facebook, not Friendster, in social networks. But in a case of the exception proving the rule, the company that broke the most ground in what would be known as ecommerce is still the company that dominates today: Amazon.

■

IN 1992, TWENTY-EIGHT-YEAR-OLD Jeff Bezos was the youngest-ever senior vice president at a Wall Street hedge fund company known as D. E. Shaw. One of Bezos's chief duties at the firm was to help launch new business initiatives. Around 1993, he was tasked with investigating the business opportunities inherent on the Internet. And among the many ideas that Bezos presented, the one that really caught Bezos's boss's fancy was the one that D. E. Shaw employees would later remember gained the nickname "everything store." The idea was simply to harness computer networks and the Internet to be a sort of intermediary between buyers and sellers of every product sold, creating efficiencies and taking a small percentage for the trouble. But Bezos quickly decided that an everything store was a bit too grandiose, and so he instead began investigating product categories that might be suitable as a proof of concept. He weighed roughly twenty different possibilities, including computer software, office supplies and CDs. He settled on books as the best test case because, as Brad Stone put it in his history of Amazon called, not coincidentally, *The Everything Store*, books were "pure commodities; a copy of a book in one store was identical to the same book carried in another, so buyers always knew what they were getting."[1] This is different than something like clothing, which has all sorts of vagaries when it comes to details like size, cut, shape, and color. Books also had an advantage over something like CDs because, at that time, there were only two major book distributors that every bookseller in the country worked with, Ingram and Baker & Taylor. This compared favorably to the several major and hundreds of minor record labels in the world. And as Stone also points out, books have what we would now refer to as a strong long tail: there were three million different titles of books in print worldwide, as opposed to only 300,000 different titles on CD.[2] No single store could shelve all those titles. But an online store could. As Bezos himself would later say, "With that huge diversity of products [titles] you could build a store online that simply could not exist in any other way."[3]

It seems that in the course of his research, Bezos, like Jim Clark, was bowled over by the sheer growth numbers he encountered. He ran across some data by an analyst who claimed that the amount of bytes transmitted over the web from January 1993 to January 1994 had increased roughly 205,700%.[4] As Bezos himself later pointed out, "Things just don't grow this fast outside of petri dishes."[5]

In the spring of 1994, Jeff Bezos left D. E. Shaw and struck out on his own to found an online bookseller. In multiple retellings of this founding story, Bezos has mythologized the moment as the classic entrepreneur's dilemma. He would be leaving a safe, lucrative career on Wall Street to go off on his own, with uncertain prospects for success. But that was okay. "I knew when I was eighty that I would never, for example, think about why I walked away from my 1994 Wall Street bonus," Bezos said later. "That kind of thing just isn't something you worry about when you're eighty years old. At the same time, I knew that I might sincerely regret not having participated in this thing called the Internet that I thought was going to be a revolutionizing event."[6]

The well-worn legend is that Jeff Bezos and his wife MacKenzie packed up their car and headed west, unsure of where they were going, with Jeff typing up a business plan on his laptop as they drove and phoning angel investors along the way on his cell phone. But the truth is, Bezos had already flown out to California to recruit software engineering talent. And according to multiple accounts, he likely knew the destination of his cross-country car trip would be Seattle. His careful research had shown him that Seattle had the advantage of being a tech hub—home to Microsoft of course, and thus filthy with tech talent—and also that it was a six-hour drive from a major warehouse that book distributor Ingram operated in Roseburg, Oregon. Also, Washington State was not nearly as populous as California. No doubt, his research had also led Bezos to realize that a company did not have to charge sales tax unless it had a physical presence in the state a customer ordered from. So, Washington being less populous than California was a major plus. Other locations Bezos considered for the benefits of tax purposes were Portland, Oregon; Boulder, Colorado; and Lake Tahoe, Nevada.

The company that would become Amazon was founded in the summer of 1994 in the garage of the home that Jeff and MacKenzie Bezos rented in Bellevue, Washington, at 10704 N.E. 28th Street.[7] Jeff and MacKenzie were the founding employees, along with a couple of programming talents that Jeff had recruited earlier. One was Shel Kaphan, who would go on to write much of the initial structure that would become the Amazon site and who many people thus think of as a cofounder of Amazon in all but title. "When I got there, [the company] was basically not even a business plan on paper," Kaphan says. "It was a couple of spreadsheets and a verbal description of [the concept]. The garage, which had been converted, was just a not particularly well-heated part of the house."[8]

As a *Star Trek* fan, Bezos originally kicked around the idea of naming his company MakeItSo.com, after Captain Picard's famous catchphrase. Relentless.com was also considered as a way to suggest that the company would be relentlessly focused on customer service. But that was rejected as sounding too menacing. For a long time, a strong contender was Cadabra, but Kaphan talked Bezos out of that name, claiming it sounded too close to cadaver. Browse.com and Bookmall.com were also rejected, as were the alphabetically advantageous Aard.com and Awake.com. Finally, Bezos himself settled on Amazon. As he would later say, "This is not only the largest river in the world, it's many times larger than the next biggest river. It blows all the other rivers away."[9] The earth's biggest river;

the earth's biggest bookstore. The domain name was registered on November 1, 1994.

■

IT'S INTERESTING TO REALIZE, given Amazon's later reputation for warehousing, logistics and fulfillment mastery, that at launch, the company didn't have the resources for a proper warehouse. Initially, Amazon would take a catalog of available books entitled Books In Print, sent out by R. R. Bowker of New Jersey, bring it online, add some search functionality and allow customers to find the books they wanted.[10] Books In Print was basically the industry bible, the source that every bookseller in the country, large and small, used to order titles. When you went to your local bookseller and asked for a specific book to be special ordered, Books In Print was the resource they referenced to see if they could accommodate you. All Amazon did was take this resource, insert itself as the middleman, and take it directly to consumers. Could R. R. Bowker have put Books In Print online itself? Probably. But it didn't, and Jeff Bezos did. Amazon supplemented this catalog with inventory data from the two major book distributors, Ingram and Baker & Taylor. When a customer searched for a book, Amazon ordered the title itself, took delivery of it temporarily, and then turned around and shipped it to the customer.

In the spring of 1995, Amazon conducted a semiprivate beta test among friends and family. Almost right away, Bezos and company discovered that the promise of "every" book in the world was enticing to people. The first orders to come in weren't for the latest bestsellers, but for obscure titles that might not be carried at your average bookstore. The first-ever order of the beta test, and thus the first-ever Amazon order, was from a former coworker of Shel Kaphan named John Wainwright, whom Kaphan had invited to the beta test. Wainwright ordered the book *Fluid Concepts and Creative Analogies* by Douglas Hofstadter on April 3, 1995.

Amazon offered the bestsellers too, of course, and heavily discounted them as loss leaders. But it would be more obscure titles like *Fluid Concepts and Creative Analogies* that would allow Amazon to create a rabid following among early adopters. The bestselling title for Amazon's first year of existence was *How to Set Up and Maintain a World Wide Web Site: The Guide for Information Providers* by Lincoln D. Stein.[11]

But obscure titles presented problems of their own. Amazon tried to deliver books to customers within a week. Rare finds could take as much as a month to track down. And even then, Amazon still had to order the books, receive them, repackage them, and send them back out to customers. Furthermore, it turned out that distributors required retailers to order a minimum of ten books at a time. During the beta, Amazon of course didn't have that sort of sales volume. "We found a loophole!" Bezos would later remember proudly. "Their systems were programmed in such a way that you didn't have to receive ten books, you only had to *order* ten books."[12] So the Amazon team found an obscure book about lichens that was listed in the system but was regularly out of stock. They began ordering the one book they wanted and nine copies of the lichens book. The book they wanted would ship while the distributor promised to track down more copies of the lichens book.

Many, if not most, of the early customers phoned in their credit card numbers, not trusting the online transactions to be safe. "Some people would even just email their full credit card number to us," says Kaphan, "as if that was somehow more secure than entering it in a form on the web."[13] To make sure that the orders were secure from hackers, credit card numbers were recorded on one computer, copied to a floppy disc and then physically walked to a second computer, which would batch the transactions. This was known within Amazon as sneakernet. The sneakernet system was eventually retired, but as Kaphan notes, "It was quite a while, actually, before we had enough business to justify a full-time connection to a credit card processor."[14]

Amazon was by no means the first ecommerce player to launch; but it had the ambition to incorporate some key innovations the web made possible, many of which we take for granted today. These innovations were meant to show that ecommerce could do things traditional commerce couldn't. For one thing, think of the basic user interface of ecommerce: the shopping cart. If users are shopping your site, they might have several things to purchase. You don't want them to have to begin the checkout process for each item they want to order. You need a virtual place to store the items customers are considering. You want a virtual shopping cart. Amazon popularized this metaphor. From a technical perspective, remembering a given customer from one visit to the next is a useful thing. Amazon remembered what a customer had ordered previously—or almost ordered, before abandoning their cart—so it could store that information and prompt the returning customer accordingly. Again, using cookies, Kaphan and his small team set up the site to change so that once a customer bought a book, it wouldn't be promoted to them again. Today we're used to the idea that when I visit an ecommerce site I might see offers for entirely different products than you might, based on our different shopping histories. Amazon was one of the first sites to tailor its storefront individually in this way.

And then there was the brilliant innovation of product reviews. Prior to the Internet, few general retailers offered reviews of the products they were selling. A supermarket doesn't say one brand of toothpaste is higher-rated by shoppers than another. Quite the opposite in fact: a traditional retailer wants to be seen as a neutral broker. But Amazon felt it needed to mimic a real-world book retailer in one key aspect: acting as a source of recommendations. So, Shel Kaphan hacked together a rudimentary rating system over a weekend and initially designed it to provide editorial content from Amazon itself. But this soon evolved into allowing reviews from anyone and everyone. User ratings and reviews were controversial, as, obviously, authors resented bad reviews getting posted prominently alongside their books on the sales page. But to its credit, Amazon stuck to its guns, believing that honest reviews, as well as a reputation for helping customers make smart purchasing decisions, would be a key differentiator compared with offline retail.

In coming years, all these innovations would combine to give birth to the famous recommendation engine. Tying in with the cookies and session ID systems, the recommendation engine would parse your own browsing history, your own purchasing history, as well as the purchasing history of everyone else on Amazon, to help give users that classic prompt: if you liked x, then you will probably like y. Today, this is a key component of not just ecommerce, but of things like Netflix and music-streaming services like Spotify. Initially for Amazon, however, it was just another differentiator from offline retail, a way to prove that ecommerce could do things traditional retail could never dream of.

■

AMAZON'S FULL WEBSITE launched to the public on July 16, 1995. The only graphics included the early Amazon logo (which was a field with a river running through it, and a giant *A*) and tiny pictures of the covers of featured books that Amazon was promoting. All books on the site were discounted by a blanket 10%, but the spotlight books were discounted 20% to 30%.

Sales were slow. Early on, a dozen purchases constituted a good day. But that was a good thing because everything was being done by hand. When an order came in, Amazon turned around and ordered the book from the distributor, who shipped the book to Amazon's meager offices. Then, the handful of Amazon employees, Bezos and Kaphan included, reboxed the books and shipped them to customers. The company had one public-facing email address and all the employees would take turns responding to customer inquiries.

Over its first week in business, Amazon rang up $12,438 worth of book sales. But it was able to ship only $846 worth out to customers.[15] By October, Amazon had its first hundred-order day. And though those numbers sound good for a business blazing an entirely new trail, the fact of the matter was, it would not be enough to sustain operations for very long. For one thing, around the time of the site launch, Amazon had moved into a larger space at 2714 First Avenue South in the SoDo neighborhood of Seattle, across the street from the headquarters of Starbucks. They were a real business now, and they were trying their best to learn to act like one, which did not come cheap.

In later SEC filings, we can see that despite steadily growing sales, by the end of 1994, Amazon lost $52,000. In its first full year of operations, 1995, Amazon was able to sell half a million dollars' worth of books, and yet it was still in the red to the tune of $303,000.[16] And that brings us to the question of financing, which, if you'll notice, we haven't really mentioned up until this point. That's because for as long as he possibly could, Bezos was determined to self-fund the business. Drawing from the money he had socked away over his years on Wall Street, as well as with a mixture of credit card loans and personal guarantees, Bezos was able to fund the company through early development. In the summer of 1995, in the name of her family trust, Jeff's mother, Jackie, invested $145,000 in the company, a literal friends-and-family round. But that wouldn't be enough to keep the lights on very much longer.

So, in the summer of 1995, Jeff Bezos started to try to raise money for Amazon for the first time. He didn't want to approach big-name venture capital firms, and instead solicited from Seattle connections he knew personally. The business plan Bezos shopped around to these local investors was projecting $74 million to $114 million in sales for Amazon by the year 2000.[17] On the strength of these projections, Bezos was able to raise $981,000 by the end of 1995, giving away around 20% of the company.[18] Of course, those investors would do quite well, because that best-case scenario they bought into was not even close to what Amazon would eventually achieve. By the year 2000, Amazon would record $1.64 billion in net sales, more than fourteen times Bezos's rosiest estimate.

The true turning point for the company came when Amazon was featured on the front page of the *Wall Street Journal* on May 16, 1996. Under the headline "Wall Street Whiz Finds Niche Selling Books on the Internet," the *Journal* described Bezos as "a whiz-kid programmer on Wall Street" who "suddenly fell under the spell of one of the iffiest business propositions of modern times: retailing on the Internet."[19] The impact of the article was instantaneous. Almost overnight, Amazon went from being a tiny curio on the corner of the Internet to becoming the standard-bearer for a whole new industry. The search engines and AOL came calling, interested in forming partnerships. Just as important, the big-name venture capital firms that Jeff Bezos had deliberately avoided until now began circling as well. Amazon held out for the crème de la crème, and successfully landed an $8 million investment from Kleiner Perkins for 13% of the company, with no less that John Doerr agreeing to sit on Amazon's board of directors.[20]

"Jeff was always an expansive thinker, but access to capital was an enabler," Doerr has said of Bezos.[20] Suddenly, a new motto was making the rounds at Amazon, a phrase that would become the standard rallying cry for every dot-com–era business: Get Big Fast. Netscape coined the term originally, but Jeff Bezos and Amazon turned it into something just short of an official motto. In essence, the initial thinking behind Get Big Fast was practical. The publicity surrounding the *Wall Street Journal* article no doubt alerted bigger competitors to Amazon's existence. Borders and Barnes & Noble were now aware of Amazon, if they hadn't been already. In the *Journal* article, it was noted that Amazon had been on track to do about $5 million in revenue that year, which represented the yearly sales of a single Barnes & Noble superstore. Bezos knew Amazon would have to do better than that, and quickly, before Barnes & Noble launched a website of its own. If the "Earth's biggest bookstore" really could go toe-to-toe with the entire book-retailing industry, it was time to put the pedal to the metal.

To this end, Bezos and Amazon began spending the recently raised capital infusion on people: warehouse staff, technical support, product reviewers, etc. So many people were brought on board so quickly that an early Amazon HR manager sent a much-remembered pronouncement to local recruiting firms to "send us your freaks," the oddballs and misfits who might not suit a typical office or typical company, but might be able to thrive in the chaos of a Get Big Fast company.

In November of 1996, Amazon moved again, into new digs in South Seattle, across the street from a pawn shop and a strip club that advertised "12 beautiful women and one ugly one."[22] This new building housed a proper distribution facility, boasting 93,000 square feet of space.[23] This move coincided with the hiring of

Oswaldo-Fernando Duenas, a 20-year veteran of FedEx who was the first person at Amazon with extensive logistics and warehousing experience. Also around this time, roughly the fall of 1996 through the spring of 1997, Amazon hired veterans of Kraft Foods and Symantec to handle marketing, an ex-Microsoft engineer, brought in to handle product development, and an executive from Barnes & Noble to head business expansion.

Barnes & Noble had certainly taken notice of what Amazon was up to. In late 1996, the Riggio brothers, Leonard and Stephen, who had built Barnes & Noble into the 466-store juggernaut that made it the Wal-Mart of the book-retailing industry, flew out to Seattle to have dinner with Bezos. According to Tom Alberg, an advisor to Bezos at the time, the Riggios said they admired what Amazon was doing, but when and if Barnes & Noble got around to selling books online, it would crush Amazon. According to Alberg, the Riggios originally wanted some vague partnership, with Len Riggio saying, "I want to invest. I want to own 20 percent of you. I don't care what the price is."[24] But Bezos didn't take the bait.

The question was, if Barnes & Noble created a website, could it do so better than Amazon? Bezos calculated that they could not. In short, he would lure the offline retailers onto a battlefield of his choosing, which was the web. He trusted that the web offered Amazon an advantage in skill sets that would prove decisive. While the offline retailers would spend millions to copy Amazon's operations online, Amazon would meanwhile be outflanking them by moving into new markets.

Barnes & Noble launched its own website on May 12, 1997, and locked up an exclusive agreement with AOL to become that service's exclusive bookseller.[25] This was back when accessing AOL's 8 million early online subscribers was invaluable for young web companies hoping to compete. And of course, Barnes & Noble attempted to leverage customer familiarity with those 600-odd physical stores scattered around the country. Very smart people looked at the competitive situation and declared that Amazon was doomed. In September of 1997, *Fortune* magazine had a story with the title "Why Barnes & Noble May Crush Amazon." In the article, the author posited, "Anything Amazon.com can do on the Internet, so, too, can Barnes & Noble."[26] Famously, Forrester Research released a report in early 1997 entitled "Amazon.toast."[27]

But Amazon fought back with the Netscape playbook. It IPOed on May 15, 1997, gaining the now-requisite flood of media attention. Shares went out at $14 to $16 per share, but closed on the first day of trading at $23.50.[28] It wasn't a mind-blowing first-day pop like Netscape or Yahoo, but investors had been intrigued by Amazon's strong growth numbers. In 1996, sales were $15.7 million. In 1997, sales would top $147 million.[29] At the time of the IPO, Amazon was recording a 900% growth in revenue.[30] The promised efficiencies of the ecommerce model that Bezos had so much faith in were actually panning out. Amazon was turning over its inventory 150 times a year; traditional physical bookstores like Barnes & Noble turned inventory only 3 or 4 times a year.[31]

Just as Bezos had anticipated, he, not the Riggios, was the incumbent on the web. Barnes & Noble had to spend tens of millions of dollars to create a website, and even after doing so, it never drew significant numbers of shoppers back from Amazon. Amazon, meanwhile, was steadily poaching customers from Barnes & Noble's website, while at the same time chipping away at offline retail sales. It wasn't clear, in fact, that having a nationwide chain of stores offered any sort of advantage whatsoever against an online insurgent. This was antithetical to everything people understood about retail sales. Being local neighbors, Howard Shultz, CEO of Starbucks, once met with Bezos to propose some sort of partnership that would allow Amazon to place merchandise in Starbucks's own stores, perhaps in a bid to emulate Barnes & Noble's cafés. Shultz told Bezos, "You have no physical presence. That is going to hold you back." Bezos shot back that physical presence wasn't necessary: "We are going to take this thing to the moon."[32]

Disruption is the word that has come into common parlance to describe the nature of these encounters and the Barnes & Noble/ Amazon battle would be the first of the great contests between online disrupter and offline incumbent. So, it's interesting to note: Amazon didn't exactly trounce its initial competition. Barnes & Noble is still around (though Borders is gone). Bookstores are still around, unlike, say, video rental stores or music stores. Amazon didn't surpass Barnes & Noble in total revenue as a company until 2004.[33] Amazon didn't even become the biggest book retailer in the world until 2007.[34] But it didn't really matter, because just as had been the plan all along, while the incumbent booksellers raced to copy Amazon, Amazon was already moving toward new horizons. Jeff Bezos didn't care if Amazon ever definitively "won" in books, frankly, because his real aim was to take increasingly bigger bites of other markets, and then other markets, and other markets, until one day, Amazon had a piece of *every* market.

It seems that at some point between researching the web at D. E. Shaw and the Kleiner Perkins investment, Jeff Bezos convinced himself ecommerce really was a markedly superior way of doing business. Like Andreessen and Clark at Netscape, Bezos saw the horizons on the Internet as being unlimited. Time and again, in several different interviews and speeches, Bezos would talk of how this was "day one" of the Internet revolution. Bezos believed Amazon had a chance to not only establish ecommerce as a viable proposition, but also to disrupt the entire system of buying and selling everything. Bezos wasn't just thinking about books, but about retail itself, a business model that went back millennia to that first day merchants gathered in a central location to hawk their goods to a local population. In Bezos's vision, the products would come to the people. First books, then anything else. In the end, he would make the everything store a reality.

As Amazon executive Joy Covey remembered, Bezos "always had a large appetite. It was just a question of staging the opportunities at the right time." Amazon launched its music store in June of 1997 and its movies store in November of 1997.[35] A mere 120 days after launching the music store, Amazon.com could claim to be the largest online seller of music. The motto on the top of the website was changed from reading Earth's Largest Bookstore to now read Books, Music and More, and eventually would simply say Earth's Biggest Selection.[36]

In the mid-nineties, a cautionary tale began to be bandied about the business world: beware because your industry could suddenly be Amazoned! No matter what you sold or what service you provided, you had to be on the lookout for a web startup (often Amazon itself) that might come along and attack your market. This upstart might seem like a tiny pretender at first, but their web magic would start wooing customers, and before you knew it, that little dot-com might have a bigger market cap than you did. A key factor that would contribute to the coming dot-com bubble would be the untold billions that companies in all industries spent in an attempt to be proactive and come up with an "Internet strategy." To avoid being Amazoned. Bezos himself would later say of his first competitor's sudden efforts to compete on the web, "Barnes & Noble isn't doing this because they wanted to. They're doing this because of us."[37]

7

TRUSTING STRANGERS

eBay, Community Sites and Portals

It's often remarked upon that Silicon Valley has a prominent utopian streak. When founders of today's billion-dollar chat apps talk earnestly about how their inventions are "changing the world," they are part of a long tradition of grandiose digital idealism indigenous to the tech industry. A lot of this comes from geography and timing. Silicon Valley came into being in the 1960s and 1970s. Cold War–era defense- and space-research spending seeded the technology industry in the Valley, while the nearby counterculture havens of Berkeley and San Francisco infused flower-power thinking among the denizens. So, Silicon Valley has always been equal parts egghead libertarianism and acid-tinged hippie romanticism. Both of these worldviews mesh quite well actually when it comes to believing that technology can be used to better mankind and free it from all manner of oppression, repression and just everyday drudgery. The Internet was another in a long line of technological miracles that many believed would elevate minds and free souls from all sorts of impediments. For the libertarians the Internet was great because it had few rules and no governance. For the hippies, the Internet promised free expression and a democratization of ideas.

Steeped in this milieu was a French-Iranian immigrant named Pierre Omidyar. Omidyar had been involved in the Silicon Valley startup scene even before the Internet Era started. When Microsoft purchased eShop, the startup he worked at, Omidyar's share of the windfall made him a millionaire. Not even thirty at this point, he had no intention of retiring. Omidyar came from the libertarian side of the Valley's intellectual duality. With that philosophical bent, he found himself wondering if perhaps the then-exploding web could be a sort of laboratory for realizing that long-held libertarian dream: a perfect, frictionless, regulation-free marketplace. His insight was that the traditional classified ad—say, selling a used coffee table by buying a few lines in the newspaper—just wasn't an efficient use of market dynamics. With a normal ad, you simply said, "I want $100 for this table." And if someone agreed that that was a fair price, then you got your $100. But what if $100 wasn't the right price? What if you could have gotten more for your coffee table? What if the buyer could have paid less? There was no way of knowing. In a perfect marketplace, the market price is the *correct* price because buyers and sellers (ideally, multiple buyers and sellers) can haggle to arrive at an optimal result. Classified ads did not allow for that haggling. But what if you could create an auction scenario in classified ads? That way you could find the true market price for any item because the buyers and sellers would arrive at the final price organically. As Omidyar described it, "If there's more than one person interested, let them fight it out. The seller would by definition get the market price for the item, whatever that might be on a particular day."[1] In other words, Omidyar didn't just want to bring classified advertising to the web; others like The Monster Board for employment classifieds and Match.com for personals were already doing that. He wanted to see if the web could create the perfect classified platform by introducing the auction element.

On the Friday night before Labor Day weekend in 1995, Omidyar holed up in his home office on the second floor of his town house and began writing code for his auction idea. By the end of the long weekend, he had cobbled together a crude website that allowed users to do three simple things: list items for sale, view items that were on sale, and place bids on those items. He hosted the site on his home server and published it to the web via his $30-a-month account with a local

ISP. He called the site AuctionWeb. But he hosted it as a subsite on his personal webpage, ebay.com. So, the URL was ebay.com/aw.

Why eBay? Well, after cashing out from the eShop sale, he had done some web consulting and freelance work and decided to do so under the rubric Echo Bay Technology Group, a name he simply liked. However, the domain EchoBay.com was taken, so he registered what he considered to be the closest approximation: eBay.com. Omidyar was already hosting an assortment of other properties on the domain, so AuctionWeb was born sandwiched between a handful of other sites, including one with links to recent Ebola outbreaks, an interest of Omidyar's.

As far as Omidyar can recall, not a single visitor came to AuctionWeb on its first day online. In order to drum up interest, he posted a message about the site on the National Center for Supercomputing Applications website—the NCSA still being a heavily trafficked destination of the web at that point. The NCSA had a "What's New" page, so Omidyar posted there, describing AuctionWeb as "The most fun buying and selling on the web."[2]

Visitors to AuctionWeb began to trickle in. Thanks to one of Omidyar's many early newsgroup postings, we can get an idea of some of the offbeat items that people were listing. On September 12, 1995, Omidyar made a post on the newsgroup misc.forsale.noncomputer, where he listed items on offer as well as their current bids. Among them: autographed Marky Mark underwear (current bid: $400), a used Toyota Tercel (current bid: $3,200) and a Mattel Nintendo Power Glove (current bid: $20).[3]

After the slow start, Omidyar himself was surprised by the way AuctionWeb began to take off. Within a month, there were entire Sun computer workstations listed for sale, and even a 35,000-square-foot warehouse in Idaho for which the bidding started at $325,000. By the end of the year, AuctionWeb would play host to more than 1,000 auctions and more than 10,000 individual bids.[4] At this point, Omidyar was still running AuctionWeb as an after-work-hours experiment, for free. Both of those arrangements couldn't last forever. Because of the increase in data he was using, his ISP contacted Omidyar in February of 1996 and told him they were jacking up his hosting fees to $250 a month, the rate for a commercial account. Omidyar objected that he wasn't actually running a commercial enterprise, but the ISP didn't believe him. So, it was at that point that Omidyar figured that if he was being treated as a commercial enterprise, he might as well just become a commercial enterprise. He made two big changes to AuctionWeb. First, he decided that buyers could continue to use the site for free; their only cost would be whatever they agreed to pay the seller for the item at auction. Second, he decreed that from then on out, sellers would have to fork over a percentage of the final sale price. That percentage was set at 5% of the sale price for items listed below $25 and 2.5% for items that sold for a price above $25. These changes were implemented based on no research or calculation whatsoever, merely Omidyar's own instincts.

Omidyar had no idea if charging a fee would bring an end to his little experiment or not. Furthermore, he had no way of actually enforcing payment. He didn't have a credit card merchant account or even a method for validating auction results. In keeping with his libertarian ethos, however, he refused to impose any governance or policing of his system. He simply relied on sellers to be honest.

It turned out that his faith in humanity was justified, because envelopes started showing up in his mailbox with checks inside them. By the end of that first month of February, when Omidyar tallied up the envelopes, he found that he had made more than the $250 he needed to cover his web hosting. And just like that, eBay became that rarest of things: the first-ever meaningfully profitable ecommerce company.

Soon AuctionWeb was more than just nominally profitable. Very quickly, it became meaningfully lucrative, especially for one man and his hobby. In March of 1996, revenues hit $1,000. In April, $2,500. And in May, $5,000. Revenues would double again in June, surpassing $10,000. Omidyar had a revelation. "I had a hobby that was making me more money than my day job," he recalled. "So I decided that it was time to quit my day job."[5]

A lot of AuctionWeb's early user growth came from things like antiques and collectibles because, unwittingly, Omidyar was tapping into something the Internet had been very good at from its inception: providing a platform for niche interests. From the very first days newsgroups and email began, geeks had been trading and selling their rare *Star Trek* memorabilia and the like. If anything, AuctionWeb wasn't bringing classifieds online so much as it was moving the ad hoc swap meets that already existed on the Usenet newsgroups and on early community websites into a centralized location.

But AuctionWeb's immediate success was also due to structural decisions that would enable the service to scale successfully. In short, Omidyar enabled AuctionWeb's community to organize itself. Early on, Omidyar listed his personal email prominently on the website. When buyers and sellers had a question or a dispute, they came to him directly. But Omidyar knew he didn't want to spend his time settling petty squabbles; his libertarian impulses told him that people should be able manage things for themselves. Oftentimes, when a buyer came to him with a complaint about a seller, he would simply forward the email along to the seller with a note that read, "You two work it out."

Another way to help the system regulate itself was the Feedback Forum. This was a public online message board where users were encouraged to leave written feedback about other buyers or sellers, in addition to a numerical rating: plus one, minus one or neutral. Once a user's rating on the feedback forum surpassed a negative four, they were banned from the site. This took the dispute resolution process out into the open and (just as important from Omidyar's point of view) out of his email inbox. The Bulletin Board accomplished this as well. It was the place where users could ask questions: "How do I upload pictures?" or "What do you think is the proper minimum bid I should set for this item?" Fellow eBay users could chime in with their input. Very quickly, as often happens in online communities, a select group of users prominently stepped forward to become regular advice gurus and trusted "experts." Omidyar had accidentally stumbled upon one of the longer-term factors in AuctionWeb's eventual success. A focus on community, on empowering the users and allowing them to function autonomously would prove to be absolutely vital.

Even as he built it to self-regulate, AuctionWeb was growing so quickly that Omidyar couldn't continue operating it as a one-man show. For one thing, he needed someone to open all the mail and deposit the checks and loose change that users were sending in. Chris Agarpao, a friend of a friend, was hired to come to Omidyar's house twice a week to open the envelopes and make the deposits. But more than that, Omidyar needed help building AuctionWeb into something more sophisticated than a hobby/experiment operating out of his spare bedroom. He would remember later, "I had a vague idea of what I needed to do as an entrepreneur, but I knew I wasn't going to be able to put together a business plan." In short, despite the fact that he was a startup veteran, Omidyar needed a "business" guy, a true partner to help run the operation.

Jeff Skoll had founded two successful companies earlier in his career, and in 1996 he found himself in California, consulting at Knight Ridder, helping the newspaper chain develop an Internet strategy beyond its Mercury Center experiment. As part of his consultancy work, Skoll was monitoring the early web to watch for threats to his employer's classified advertising cash cow. When he stumbled upon AuctionWeb, Skoll could see exactly the threat Knight Ridder was worried about. Instead of trying to help the newspapers beat back the disruption that he could see would soon come from the Internet, Skoll decided to join the disruptor, joining AuctionWeb in August of 1996.

Skoll pitched in at first by helping the company land space in an office park at 2005 Hamilton Avenue in Campbell, California. Skoll also convinced Omidyar to move AuctionWeb from the subdomain to the main ebay.com site. The Ebola site and the other subsites were removed. The service would eventually be known simply as eBay.

It was also Skoll who recruited Mary Lou Song to the company. Song, more than anyone else, would be instrumental in developing and cultivating the community that would be key to eBay's success. Song was skeptical of eBay's business model at first, and was perhaps even more dubious when she showed up for her first day of work in October of 1996. She was given a card table for a desk and a folding chair to sit on. Her office was between Omidyar's—who was seemingly always busy crunching out code to keep the site from crashing—and Skoll's, who was working on eBay's nascent business plan. Outside her office was Chris Agarpao's card table, where he was busy plowing through envelopes of checks from auctioneers.

Wary as she might have been, Song understood right away that eBay was a new type of business that had never existed before—indeed, *could not* have existed without the web. eBay was online commerce, but not in the way that Amazon was; it was a platform, but not like the operating system or the browser were. eBay was nothing more than a virtual marketplace, and by being virtual, it didn't actually *do* anything other than facilitate the interactions between buyers and sellers. It didn't store goods. It didn't ship goods. It didn't even guarantee the exchange of goods between buyers and sellers! The *one* truly tangible thing that eBay had was the goodwill of those buyers and sellers and the community they were creating—on their own—to make the buying and selling happen. eBay would be one of the first web companies to understand that all the value of its service came from the users and their community. eBay's *only* asset, in fact, was its users, and therefore the only important thing for the company to do was to make sure the buyers and sellers were happy so that they would keep coming back.

Song carved out her own role as eBay's community liaison/manager. She always referred to users as "the community," not as customers. She reached out to the de facto user-leaders who had risen organically on the bulletin boards and hired them to formally take over the task they were already performing gratis: policing the auctions and handling customer service. She also enhanced and expanded the existing community guidelines and processes for which Omidyar had laid the foundation. And it was Song who helped build out the user-reputation systems that were becoming so important for eBay's buyers and sellers. It was these systems that would soon become eBay's most valuable feature.

A new user to eBay might (rightly) be wary about buying something online, sight unseen, from a complete stranger who was hiding behind a username. If you were a buyer, how could you be sure the seller would actually send the item you paid for? Conversely, how could a seller be sure a buyer would pay up? Buyer- and seller-reputation ratings helped assuage these fears. The higher-rated a seller was, the more trustworthy they must be, right? And the mechanism functioned the same way in reverse: sellers wouldn't sell to users who, the ratings revealed, made a habit of stiffing other auctioneers. Thanks to Song's tinkering, the feedback scores eventually manifested themselves as actual numbers that got attached to a user and their auctions on the site. So, if someone was considering bidding on an auction from someone with a +48 rating, they could reasonably assume that seller had completed 48 successful auctions with satisfied buyers. Plus, buyers and sellers alike knew that if they had a bad auction experience, there was recourse: you could give the offending user a bad rating and thereby damage their reputation on the market. Everyone on eBay had real incentive to give constructive feedback. Things like fraud and serious disputes, while never 100% absent, were kept to a manageable minority of auctions.

This is a key evolution. In so many ways, over the last twenty years, the web and the Internet have slowly trained all of us to get comfortable interacting with crowds and, often, crowds of strangers. eBay was one of the first websites to show that a largely anonymous community, carefully constrained by a few guidelines and regulations, but invested in a system of online reputation, could actually work. Today, this key ingredient of ratings and reputation continues on sites like Yelp and Reddit—and especially on sites like Uber and Airbnb. It's hard to imagine that the current sharing economy could even exist without the reputation template that eBay pioneered.

When Mary Lou Song joined the company in the fall of 1996, eBay hosted only about 28,000 auctions a month.[6] After what was known within the company as the great eBay flood, in January 1997, eBay would host 200,000 auctions in that month alone.[7] As they got deeper into the first quarter of the year, eBay's brain trust realized that the site was on pace to take in $4.3 million for hosting all these new sales. AuctionWeb/eBay had made just $350,000 in all of 1996. They were on track for an astounding annual growth rate of 1,200%.[8]

There were several factors leading to this explosion in growth. For one thing, eBay noticed the power of Januarys: they came after the holiday season. That meant millions of people with millions of unwanted gifts. eBay to the rescue. But the site was also benefiting from the phenomenon Omidyar had discovered earlier: the Internet as a place where people of like interests, no matter how obscure or remote, could congregate. Suddenly, eBay was a central place where all these disparate communities of interest could find each other when they wanted to perform the fundamental acts of hobbyists everywhere: trading and collecting. Baseball cards. Barbie dolls. Postage stamps. Buffalo nickels. Quilts. Antiques of all stripes. Anything collectible. eBay became, overnight, the world's greatest flea market/garage sale/bazaar. In AuctionWeb's earliest months, the majority of the listings were for computer items and electronics. But at the beginning of 1997, antiques and collectibles suddenly rose to become 80% of eBay's offerings.[9] eBay would also piggyback on many of the hottest fads in collectibles, of which there were quite a few in the late 1990s. Furbies. Tickle Me Elmos. Tamagotchi. But the greatest of these was the Beanie Baby craze of roughly 1996 to 1999, exactly mirroring the rise of eBay.

Beanie Babies were stuffed animals developed by an independent toy manufacturer from suburban Chicago, Ty Inc. From initial animals like Flash the Dolphin, Patti the Platypus and others, Ty gradually ramped up its lineup of characters to encourage a habit of collectibility. But Ty also introduced a brilliant complication: artificial scarcity. Beanie Baby characters were not distributed to retailers equally. Part of the fun of Beanie Baby collecting was hunting down obscure characters in order to complete your collection. When, in 1996, Ty began "retiring" individual Beanie Baby models, this set off a collecting frenzy. Once, say, Buzz the Bee was sold out, the only way collectors would be able to obtain discontinued Buzz was on the secondary market—just the sort of market eBay provided.

In April of 1997, listings of Beanie Babies surged to 2,500 separate auctions, and eBay assigned them their own category. When rare and discontinued Beanie Babies suddenly started going for hundreds, even thousands, of dollars at auction, eBay reaped the attendant press attention thanks to its position at ground zero of the craze. Within a month, that single Beanie Baby category was responsible for 6.6% of the entire site's sales volume.[10] eBay was not exactly the company that Beanie Babies built, but Beanie Babies certainly brought eBay to the world's attention.

eBay was perfect for collectibles. By creating a centralized clearinghouse of hard-to-find items, it could eliminate many market inefficiencies that had existed for years. There are plenty of articles from the late nineties about hordes of eBay-ers descending upon flea markets and antiques shops around the country, scooping up virtually everything on hand in hopes of turning around and fetching higher prices on eBay. An antiques store in Maine put an old-fashioned calculator it had lying around up on eBay for $100. Once calculator enthusiasts discovered the listing, they bid the price up to $6,500. The store didn't know what it had on its hands until they put it on eBay, where the perfect buyer could discover it.[11]

This very rapidly led to the phenomenon of people building true small businesses on top of eBay's marketplace platform. Most small sellers on eBay were what they'd always been: hobbyists and part-timers who sold spare items for a little supplemental income. But in due course, perhaps tens of thousands of people came to make their entire living on eBay, some creating businesses large enough to employ dozens of people and gross into the millions of dollars. eBay was creating not just the world's largest virtual marketplace, but also the first marketplace that could rival the real world. Just as the Internet allowed people to connect to the entire world, eBay allowed a person to sell to the entire world from their tiny little corner of it.

And eBay embraced its image as the hobbyists' mecca. Many people are familiar with eBay's founding myth: how Pierre Omidyar created the site so his fiancée could expand her Pez dispenser collection. But like many company creation stories, the Pez story is a fiction. The Pez story was created by Mary Lou Song to get reporters interested in covering eBay's role in the collectibles phenomenon. As she put it later, "Nobody wants to hear about a thirty-year-old genius who wanted to create a perfect market. They want to hear he did it for his fiancée."[12]

■

BEFORE LONG, EBAY'S VERY SUCCESS—user and auction numbers were sometimes doubling from one month to the next—became a serious problem. Omidyar's original code, which had been strung together as an experiment, proved too weak to handle the growing user base. "It was like holding back a hurricane," Song said of the surge in users over the course of 1997–98.[13]

Knowing that they needed the resources to stay on top of growth, Omidyar and Skoll decided the time had come to raise some capital. They hadn't needed to do so before, because ever since that first month Omidyar had introduced auction fees, the site had been profitably self-sustaining. Jeff Skoll returned to his newspaper industry contacts and received interest from his old associates at Knight Ridder, as well as at Times Mirror. But both companies were put off by the valuation Skoll put on eBay: $40 million. Forty million might not seem insane to modern eyes—especially for a company growing by double-digit percentage points each month and with gross margins above 80%,[14] but as Mark Del Vecchio, a Times Mirror executive, recalled later, his bosses simply couldn't wrap their mind around the very concept of what eBay was. "They kept saying, 'They don't own anything,'" said Del Vecchio. "'They don't have any buildings, they don't have any trucks.'" So, both companies passed.

eBay instead found joy by going the technology VC route. In June 1997, Benchmark Capital paid $5 million for 21.5% of eBay. By various measures, this deal would go down in history as one of the greatest investment home runs of all time. Benchmark's stake in eBay would eventually be worth $4 billion.[15] Benchmark's money came with strong suggestions that more serious management be brought in to eBay. The days of card-table desks were over. Both Omidyar and Skoll were sanguine about this, with Omidyar saying, "We were entrepreneurs and that was good up to a certain stage. But we didn't have the experience to take the company to the next level."[16] And so, a world-class manager was recruited in the person of Meg Whitman. Whitman had nothing in the way of a technical background, but she did have experience with brands and marketing. With a degree in economics from Princeton and an M.B.A. from Harvard, like Steve Case, Whitman had done a stint at Procter & Gamble, as well as Disney and the toy company Hasbro. She proved to be a perfect choice, capable of shepherding eBay into an era when it was turning into a marketplace for every brand and product category under the sun.

Whitman came on board as eBay's CEO on February 1, 1998. By that point, eBay had only 500,000 registered users. But those users exchanged more than $100 million in goods in the first quarter of 1998, generating $3 million in revenue every month. Only one quarter later, in June 1998, eBay would announce its one-millionth user. When eBay went public on September 21, 1998, its stock popped 197% on the offer price. The company was valued at almost $2 billion. Nineteen ninety-eight was, as we'll see, the year that the dot-com mania really struck, and eBay would become one of the true highfliers of the era. Roughly two-thirds of the pre-IPO staff—about seventy-five people—became paper millionaires at eBay. By July of 1999, *Forbes* magazine would peg Pierre Omidyar's eBay fortune at $10.1 billion, Jeff Skoll's at $4.8 billion and Meg Whitman's at about $1 billion.

■

THE INTERNET ERA might have been launched in Silicon Valley, but to a large extent, it was monetized by startups in New York City. As the web began to call out for digital advertising as a revenue engine, young New York–based geeks stepped up to create digital agencies, brokerages and advertising companies. There was a new technology on the scene. The olds couldn't quite grok it, so they turned to the youth to bring them up to speed. The phenomenon of young interns being summoned to the executive suite to give presentations on the new digital realities became common. "We were all twentysomethings in really bad suits," remembered Seth Goldstein, founder of one of the first New York–based Internet marketing firms, SiteSpecific.[17] But they seemingly had a grasp on the future, so the usual rules of decorum and seniority were increasingly overlooked.

The young techies on the East Coast had a sense of fearlessness and a DIY ethos that was possibly more aggressive than even the moxie displayed by their peers in Silicon Valley. A perfect example is Craig Kanarick who, fresh off his efforts designing the first banner ads for AT&T and *HotWired*, founded the interactive media and advertising agency Razorfish with his childhood friend Jeff Dachis. The two twenty-somethings ran their "company" out of Dachis's Alphabet City apartment and suddenly found themselves consulting with Fortune 500 companies like Time Warner for no other reason than that they claimed to "get" the web. SiteSpecific had its first offices in a "hovel" on Broadway just north of Madison Square Park. The startup made great efforts to hide its squalid condition and give off an air of professionalism to their old-media clients. "Seth would call all his friends and say, 'Come in and look like you're working,'" remembered SiteSpecific co-founder Jeremy Haft. "So we would all arrive fifteen minutes before the client would arrive, and would be sitting at our desks typing away. You put up shadow puppets, and 'Look! We're a company!'"[18]

For various reasons, the design, marketing and advertising startups sprang up around, and especially below, Madison Square Park and the Flatiron Building. This entrepreneurial "scene" acquired the nickname Silicon Alley, a sobriquet that many people claim credit for but which owes its popularization primarily to New York–based advertising startup DoubleClick. DoubleClick, founded by Kevin O'Connor and Dwight Merriman, would create the first large-scale advertising network and marketplace on the web, brokering and delivering the banner ads that would generate revenue for many of the advertising-supported websites in the late 1990s. By 1998, DoubleClick was serving up more than 1.5 billion ads a month and had one of the first significant Silicon Alley IPOs in February of 1998.[19] Flush with success, DoubleClick hung a banner behind the Flatiron Building that, at the height of the company's success, declared to the world DOUBLECLICK WELCOMES YOU TO SILICON ALLEY.[20]

If Silicon Valley had a software engineering culture, Silicon Alley had a creative culture. A media culture. The DIY New York spirit spread to journalists and writers who figured the web allowed them to start publications with a global reach that could match any print publisher in the world. The best example of the web-based "ezines" that sprang up was *Feed* magazine, aka Feedmag.com, or, simply, *Feed*. Launched by two young freelance writers, Stefanie Syman and Steven Johnson, the lure was the same for independent publishers as it had been for bigger names like Time Warner: the promise of seemingly insignificant production costs. Syman and Johnson began reaching out to big names in media and culture for interviews and profiles. "And we'd be like, 'Hi! We just started this online magazine. Would you come and have a dialog about this topic?'" Syman recalled. "And they'd say yes! And we were always shocked! We were like, 'We're no one! We're not the *New York Times*, we're not *Esquire*, we're not even *Wired*.' And yet, people wanted to participate."[21]

But by then, of course, even big names like the *New York Times* were participating as well. The *Times* had experimented with a cobranded news presence on America Online called @times back in 1994. A full website went live at www.nytimes.com on January 22, 1996, with headlines, stories and pictures from the print edition. The *Times*'s local rival, the *Wall Street Journal*, limited its content solely to paying subscribers when it launched on the web in 1996. A paywall ended up being successful for the *Journal*, which eventually accumulated around a million online subscribers, proving that there were *some* types of content that audiences were willing to pay for. But time and again, publishers that went the subscriber route found that by doing so, they only left the door open for free, advertising-supported online competitors. Larry Kramer, a longtime veteran of the newspaper industry, saw just such an opportunity to deliver financial content thanks to the *Journal*'s paywall. "I said, 'I can replicate information about the stocks [investors] care about, for free on the web!'" Kramer says. "I can build a newsroom that gives them their version of the *Wall Street Journal* and the Bloomberg terminal."[22] And he did so, launching Marketwatch.com (later, CBS Marketwatch), which would IPO and earn a billion-dollar valuation, before eventually being purchased by Dow Jones, the parent company of the *Wall Street Journal* itself.

The biggest lesson to learn about online media was the 24/7 nature of the beast. The tragic 1997 death of Britain's Princess Diana was the media sensation of its day, and not just for traditional outlets like the *Times*. Online news sites like Pathfinder saw their traffic numbers spike as distraught readers went online to absorb any and all details they could find. Furthermore, web users found online forums and message boards the perfect venues to express their feelings and share their collective grief. One site that did *not* benefit from this spike in traffic was *Slate*, which had followed a long-standing publisher's tradition of taking a vacation during the summer, considered to be a "slow" period for breaking news.* And so, the whole week surrounding the Diana tragedy, *Slate* was dark, with no new content for news-hungry readers. "Diana's death finally made us understand that online journalism is by nature a round-the-clock business," *Slate*'s David Plotz would admit later.[23]

The site that best exemplified the new metabolism of media in an online environment was Suck.com. Two *HotWired* staffers, Joey Anuff and Carl Steadman, launched Suck on *Wired*'s servers in August of 1995; it was just that nobody knew it at the time. Steadman and Anuff, and eventually other *Wired* employees and outside freelancers who were let in on the secret, all published under pseudonyms. The site looked different right away. Most early websites had some sort of landing page, and usually a navigation menu, a table-of-contents–style holdover from the print paradigm to help readers get oriented. Suck completely eschewed this convention and simply put its content right there on the front page. No need to click anywhere. Suck had a simple one-column structure with reverse-chronological formatting: the newest stuff on the top, older stuff on the bottom, very much in the style of what we would later

* Weekly magazine publishers traditionally published only fifty or even forty-eight issues a year, allowing for "off weeks" around the holidays and during the traditionally news-slow summer months.

call blogs or a social networking newsfeed. And unlike any of the other sites at the time, Suck was always updating. There were no "issues" as at *Slate*. Suck tried to put up new content every day. Steadman and Anuff figured that *they* were going in to work every day and consuming content on the web in between doing their jobs (for most people at this time, the fastest and most reliable Internet connection available was often found at work), so Suck should regularly have fresh content to serve this audience of bored office drones.

The voice of Suck was pitched to people just like them: jaded cubicle warriors, Gen Xers, grunts in this new web revolution. Suck was not stentorian, like traditional media. It was first-person, confrontational, skeptical. The very first post was about the nascent Kurt Cobain death conspiracy culture. Another early post poked fun at Netscape's Marc Andreessen. There were no sacred cows, even among the digerati. But the Sucksters reserved their most cutting missives for digital Luddites. Here's a quote form a typical post. The pseudonymous author "Pop" describes his frustration with the clueless suits he is forced to work for in the new media world:

> They don't browse. They don't keep up. They read about the web, fer chrissakes, in the New York Times and in the Wall Street Journal. They tell their flunkies to order up some presence and have no idea what they've done or what it should look like. They're virgins who've been told about sex and think they have a clue. They're experts vicariously.

The columns, posts and diaries sometimes followed a regular topic or subject matter. Sometimes they were just random screeds. Some posts were well researched, almost "serious" journalism. Just as often, they were just gossip items or analysis of web industry news. And this was Suck's crucial contribution: a lot of the formal structure and stuffy posture of "traditional" media writing was abandoned. The posts on Suck always felt like they came from a distinctly personal point of view. There was commentary, sometimes overt, but also between the lines. Suck was rude, often crude, glib and satirical, but always with purpose. Suck was, in short, snarky. It was a publication that laid the groundwork for blogging in its modern form, both in structure and in tone.

■

BY 1996 AND 1997, AOL was consolidating its position as perhaps the dominant player in the new Internet economy, surpassing 10 million subscribers in 1997.[24] To serve this audience, longtime AOL executive Ted Leonsis was tasked with creating AOL-specific content that would extend the AOL experience and allow the online service to compete with what the web had to offer. Under initiatives variously called AOL Studios and AOL Greenhouse, Leonsis began to shepherd new sites into existence, often on AOL's proprietary pages, but also with experimental web presences as a way of hedging AOL's bets. There were sites devoted to fitness (The Health Zone), golfing (I Golf), finance (The Motley Fool) and people of color (Net Noir).

Candice Carpenter Olsen, Nancy Evans and Robert Levitan were media veterans who were consulting with Leonsis to develop Greenhouse sites. AOL had noticed that, for the first time, women—especially stay-at-home mothers—were beginning to come online in big numbers. So, Leonsis commissioned the trio to create a parenting-focused site called Parent Soup. With their background largely in publishing, Parent Soup launched with a magazine mindset and a plethora of professionally written articles and parenting advice. But right away, it became obvious that what the users really liked were the message boards around the articles, where they could trade tips, experiences and stories with other users. "Once they came in, yeah, they read the content," Evans says. "But the content was the appetizer. They congregated at the message boards. They began talking to each other. I remember this one mother going, 'I am just so thrilled to be talking to someone today who could talk in complete sentences.'"[25]

Expanding on this lesson, AOL funded a standalone website targeting the female audience more generally, eventually called iVillage. Content was still a key component for drawing users in, but iVillage consciously focused on the message boards and forums as well. Again, the lesson was that users in an online community were perfectly capable of producing value all by themselves. The community aspect of sites like iVillage became more than simple chatting and interaction, it became a way for people to live their lives online. "'iVillage got me through my pregnancy, iVillage got me through my breast cancer, iVillage got me through my divorce,'" says Evans. "It was all those women together. Women *got* the webbiness of the web. The web was made for them."[26]

Like eBay had done, a growing crop of community-based sites realized that their most valuable asset was their users. Today, we take for granted that social-networking sites like Facebook are merely platforms for user activity. Facebook doesn't actually generate anything itself. We do. The users generate content for Facebook to advertise against. The early social sites stumbled upon this miraculous business model almost a decade before Mark Zuckerberg did.

A Los Angeles–based entrepreneur named David Bohnett started a small firm that designed and hosted websites for local businesses. In order to drum up more clientele, he hit upon the idea of giving away limited homepages to individuals for free. "I was a passionate advocate of the validity of user-generated content," Bohnett says. "That the Internet was all about giving people the opportunity to contribute and participate, and feel like they were a part of the medium—that it was not a top-down, programmed model like radio and television."[27] Bohnett provided templates and plug-and-play tools that allowed a user to create a rudimentary homepage without having to know HTML or how to find a host or a server. Bohnett's brainstorm was to group the homepages into groups of similar interest using a virtual real estate model. So, you could homestead your website in a "neighborhood." For example, Nashville for country music sites, Area 51 for science and technology, or West Hollywood for LGBT sites.

GeoCities, as the site was called, proved to be wildly successful by pursuing the "let a thousand flowers bloom" strategy to its conceptual extreme. Millions of GeoCities homepages were created, often by individuals, with most being nothing more than simple personal pages with variations of a "Hello World" message. Similar plug-and-play homepage hosts sprang up called Tripod and Angelfire, both allowing users to express themselves directly by producing rudimentary "profiles." GeoCities and the like were "social media," or at least, an early form of it. What they weren't, *precisely*, was "social networking" because despite the fact that GeoCities grouped like interests together, the focus was not *exactly* on mapping social connections. Not yet.

If Bohnett eschewed the "top-down" model of media, other entrepreneurs thought that the web itself could be a powerful new model of top-down media, at least in the broadcasting sense. Mark Cuban was a retired entrepreneur who had made his millions selling a company to CompuServe in 1990. As the web was taking off, Cuban was approached by a college acquaintance from his alma mater, the University of Indiana. "There's gotta be a way that we can listen to Indiana basketball even if we're in Dallas," Todd Wagner told Cuban.[28] The pair formed AudioNet, which was eventually renamed Broadcast.com, in September 1995, based on that one simple premise: giving people access to streaming radio and video content anywhere in the world, via a web browser. Soon, the site was hosting 400 live events a day and was being accessed by half a million viewers daily.[29]

Cuban had the same intuition that Suck.com had: that because people were tied to their computers at work, there was a certain "prime time" for content during the day. "We reach people where they are," Cuban told *Fast Company*. "We reach more white-collar office workers during business hours than ABC, NBC and CBS combined."[30] Broadcast.com would air literally anything, even live police scanners. But it also signed exclusive deals to webcast live programming from hundreds of local radio and TV stations as well as sporting events from Major League Baseball, the NCAA and the NHL. Broadcast.com even had some community elements like SportsWorld.com, where fans could discuss the live events they were watching along with other fans.

Broadcast.com proved that just allowing people to *use the web* could be an incredibly successful business model all by itself. Sometime in 1995, two low-level Apple employees named Sabeer Bhatia and Jack Smith took this idea even further. In the mid-nineties, your email address was something that was assigned to you by your Internet service provider, by your employer at work or by your university if you were at school. And you could access your email through that provider only. Today we are used to free, almost disposable email addresses; but in the early days of the Internet, email addresses were actually something of a scarce commodity. Bhatia and Smith's idea would change all that, allowing people to check their email anywhere—at work, at home, on the road—anywhere there was a web browser and Internet access. They wanted to let users pick their own email address. They wanted to enable people to separate their personal lives from their professional lives, at least in the realm of email.

So good was this idea, and so mind-blowingly obvious was it to Bhatia, that when Smith first called on his cell phone to suggest the concept, Bhatia told him, "Call me back on a secure line when you get to your house! We don't want anyone to overhear!"[31] Bhatia wrote up a business plan for the idea, but refused to make copies for fear someone else would beat them to the punch. When Bhatia made the rounds at venture capital firms, he pitched a dummy startup concept instead of the web-based email idea. If the VCs in question rejected the dummy startup for what Bhatia considered to be the right reasons, only then would he share with them his real idea: a simple, seemingly obvious concept that would be called Hotmail.

Hotmail.com launched on the web on July 4, 1996. In little more than a year and a half, Hotmail would claim 25 million users.[32] At the time, this meant that Hotmail was actually the fastest-growing web thing in history. Such phenomenal growth was the result of a clever marketing tactic. Every time a user sent an email using Hotmail's free web mail accounts, a small link was appended at the bottom that read: "*Hotmail: Free, trusted and rich email service. Get it now.*" So, every time an email was sent, the sender was promoting Hotmail's service. The very act of using Hotmail helped spread the word about Hotmail. This kind of practice is now called viral marketing, the technique of promotion by rabid user word of mouth.

Today, this is the very foundation of modern marketing strategy; in Hotmail's era it was very much new and revolutionary.

Almost everyone on the web thought Hotmail was a brilliant idea as well. Yahoo came calling, and almost every other player in technology was interested in getting a piece of Hotmail and its viral growth. But all lost out to Microsoft, who, on New Year's Eve 1997, purchased Hotmail for $400 million in stock. Not bad for two years of work, and an idea that even its founders thought was so obvious that anyone could have done it.

■

HOTMAIL'S TIMING WAS impeccable. By late 1997, and especially through the whole of 1998, there was a big new watchword among Internet players: portal. The major search sites—Lycos, Infoseek, and especially the two most popular search destinations, Excite and Yahoo—were regularly among the most trafficked destinations on the web. And by 1997, having a lot of web traffic meant you could generate quite a lot of revenue. Yahoo, in particular, hit a seemingly insane metric: 1 billion pageviews a month.[33] And of course those pageviews translated into "impressions" for advertisers and their banner ads.

The need to produce more impressions began to change the calculus at the search sites. Yahoo, for example, had once been happy to send surfers out to their intended destinations on the web. But now all those advertising dollars were making Jerry Yang and company think twice. Money would only keep rolling in if Yahoo kept web browsers returning again, and again, and again. Suddenly, sending users off to the larger Internet wasn't as attractive as keeping them reloading Yahoo's own pages throughout the day in order to generate new ad impressions. As Yang told a television interviewer, Yahoo was facing a dilemma. "You're a search engine—once they've done the searching, why do they need you?"[34] Yahoo needed to find a way to keep users on its pages. To use a watchword that was ubiquitous at the time, Yahoo needed to get more "sticky."

To that end, Yahoo and the other search sites began to try anything that might encourage users to return habitually. First, the search sites copied the model of newspapers: they added things like horoscopes, weather reports and stock quotes. Then they realized that features like classified ads were cheap to put up and could quickly generate listing fees with practically zero investment. And if they offered, say, airline listings, the search portals discovered they could collect lucrative promotional fees as, obviously, Expedia and Travelocity would engage in a bidding war to get on their pages.

The search sites began to accumulate a utility belt of services to keep users hooked on their offerings. Things like free, web-based email, calendars, and address books, proved to be the most sticky tools of all. Once web users locked into a given portal and began to rely on one particular site for their personal email, for their scheduling, for the most intimate details of their lives, portals locked these users to repeat visits. A portal was now where you returned to again and again throughout the day, not just to search, but to manage your life.

Providing these personal services had an added benefit. Users had to "register," i.e., identify themselves. Users who registered on a portal proved to be more lucrative than the randoms who came by just to perform a search. Registered users of what became known as My Yahoo generated, on average, 238 pageviews per person, versus 58 pages for an unregistered Yahoo browser, and 3.82 hours per month on the site, versus 0.76 hours per month for someone who just came to search.[35] *And*, registration allowed the portals to charge more to advertisers. Once you identified yourself to your portal of choice in order to claim your "excite.com" email address, the site now knew your name, your general geographic location, your age, your sex, and tons of individual preferences. Sure, the portals claimed that all of this was in the interest of providing useful info like local weather conditions, personalized headlines and stock quotes. But the reality, of course, was that they now had the holy grail of marketing: demographic data to target ads against. This served to turbo-boost the advertising revenues the search sites were already generating.

Today—however uneasily—it seems we've accepted the notion

that "free" web services make their money by whoring out our personal information to marketers and advertisers. But this practice really began in earnest with the portals, which claimed they were only interested in delivering us, say, personalized sports scores for our favorite teams. All the major search sites quickly pivoted to this new portal and personalize strategy, and to say it was lucrative would be an understatement. Excite saw its revenues jump 709% in 1997 alone.[36] The four biggest search sites, Yahoo, Excite, Lycos and Infoseek, all saw their share prices increase an average of 390% over the course of 1998.[37]

All of these various players, as they feverishly pieced together features to compete in what were called the "portal wars," went a long way to creating the competitive froth that would set the stage for the dot-com bubble. Before dot-com IPOs were an everyday occurrence, the portals, with their ballooning stock prices, were able to fork over big money (at least on paper) to construct their arsenal of user features. Yahoo had wanted Hotmail first, but since Microsoft had won that battle, it made do with the purchase of a Hotmail competitor, RocketMail, for a comparatively cheap $94 million. RocketMail was quickly rebranded as Yahoo Mail.[38] Joe Beninato, the founder of an online calendar startup called When.com, took a meeting with Yahoo, hoping to get a distribution partnership. Before he could even make his pitch, the discussion turned to Yahoo purchasing When.com. This struck Beninato as a bit nutty since When.com had not even launched to the public yet. "We didn't really have anything," Beninato recalled. "We were a couple of months old."[39] Yahoo didn't end up buying When.com, but AOL eventually did. For $225 million.

The portals wanted to be all things to all people, and so they ventured into any adjacent areas that might prove lucrative. That inevitably led to experiments with ecommerce, as the portals looked jealously at the revenue that growth sites like Amazon.com were enjoying. In addition to the dozens of promotional partnerships Yahoo signed with select retailers, the company began offering its own version of an online mall, dubbed Yahoo Shopping. In order to make it easier for small merchants to set up shop in its mall, Yahoo purchased a company called Viaweb from a young British programmer named Paul Graham. By the holiday season of 1998, more than 3,000 different storefronts had opened shop, with Yahoo raking in monthly fees and a percentage of every sale.[40]

"We began with simple searching," Yang told *Time*, beginning to sound a bit like a studio mogul, "and that's still a big hit—our *Seinfeld* if you will—but we've also tried to develop a must-see-TV lineup: Yahoo Finance, Yahoo Chat, Yahoo Mail. We think of ourselves as a media network these days."[41] A Wall Street analyst told *Businessweek*, "You have to look at it [Yahoo] as the new media company of the 21st century."[42]

BLOWING BUBBLES

The Dot-com Era

For people of a certain age (my grandparents, for example), the Great Depression was not just a historical event. It was an economic and social apocalypse that, simply by having occurred once, could, ipso facto, recur at any time. It played on their minds like a psychic bogeyman. Anytime things "got too good," that could only mean a crash was around the corner. In many ways, the dot-com bubble and its subsequent bursting are a similar bogeyman, at least to Silicon Valley. Any time a new technology leads to the proliferation of startups, any time venture capital investments increase year over year, any time company valuations pass stratospheric levels and high-profile IPOs hit the market, people inside and outside of tech fall all over themselves to declare that a new bubble is here, and everyone should head for the hills. But the fact is, the dot-com bubble was a truly singular event, brought on by a unique mixture of causes, and we are unlikely to see its kind again in our lifetimes.

■

FRIDAY, AUGUST 13, 1982, might not sound like an important day in history, but in the annals of finance, it is one of the more momentous. That afternoon, the Dow Jones Industrial Average closed at

788.05, up 11.13 points, or 1.4% from the previous day's close of 776.92. The Dow would never again close as low as 776. By the end of 1982, it would cross 1,000, and in a few years, Friday the 13th of August 1982 would come be recognized as the beginning of the greatest bull market in American history. By the time the dot-com bubble burst in March 2000, the Dow and the S&P 500 Index would have risen tenfold, and the technology-heavy Nasdaq index nearly thirtyfold.[1]

There were some quite notable hiccups along the way, but from 1982 until the turn of the century, the market closed up, year-on-year, almost every single year. Even after the Black Monday crash in 1987, when the Dow lost 22% in a single day, investors who held on through the crash had more money on December 31, 1987, than they had on January 1, 1987. An entire generation of investors came of age believing that markets only moved in one direction: upward. If history tells us anything, it's that when people come to believe only good news can ever happen, a speculative financial bubble is probably inevitable. The dot-com era was really the culmination—the euphoric end-stage—of this protracted bull market.

It was all that much more impactful because it happened to the baby boomers, the megageneration. Between 1946 and 1964, 76 million Americans were born, and by the 1990s, this cohort was entering its forties, the time that most people begin saving for retirement. If the baby boomers were now interested in investing, that meant America was now interested in investing. The sheer weight of their numbers, backed by the accumulated wealth from their prime earning years, meant that there was suddenly a mountain of money looking for a place to go.

Boomers were managing their own retirement savings in much larger numbers than the generation before them, who relied on pensions rather than 401(k)s. And they hadn't grown up with the fear of the stock market crashing and causing an economic crisis. The economist John Kenneth Galbraith described just this sort of generational turnover in investing philosophy in his book *A Short History of Financial Euphoria*. "For practical purposes," Galbraith wrote, "the financial memory should be assumed to last, at a maximum, no more than 20 years. This is normally the time it takes for the recollection of one disaster to be erased and for some variant on previous dementia to come forward to capture the financial mind. It is also the time generally required for a new generation to enter the scene, impressed, as had been its predecessors, with its own innovative genius."[2]

The dot-com bubble is called the dot-com bubble because of the hundreds of new technology stocks that debuted in the late 1990s, but the fact is, the party had been going for quite a while already. From the 1987 Black Monday crash to the inauguration of President Bill Clinton, the stock market had nearly doubled. In 1995, the S&P 500 Index returned 37.20% in a single year. When the dot-com companies announced their arrival with Netscape's spectacular IPO in August of 1995, Wall Street was already in an ebullient mood. "The dot-com stocks were the froth in the cappuccino," former *Barron's* financial journalist Maggie Mahar says.[3]

Even though companies like Yahoo, Amazon, eBay and others were formed largely in the two years between 1994 and 1996 (and generally went public in the two years after that), it wasn't until 1998 that the stock prices of dot-com companies began to demand attention. It took a while for dot-com stocks to stand out because, again, at the time, seemingly all of Wall Street was doing well. Everything was already inflated. A traditional old-economy stock like General Electric was trading at forty times earnings.[4] During the time period from Netscape's IPO in August of 1995 to the beginning of 1999, shares of traditional blue-chip companies like, say, Procter & Gamble, doubled. Not a bad return in only forty months. So, at first, Internet stocks didn't seem all that exceptional.

But if you weren't content with merely doubling your money on a solid, staid stock like Procter & Gamble, then, by 1998, you might start to look enviously at the returns tech stocks were ringing up. Everything changed over the course of 1998. If you bought $1,000 worth of Yahoo and Amazon each at the time of their IPOs, over the course of 1998—merely twelve more calendar months—you would ring in the new year of 1999 to discover that your original $1,000 investment in Amazon was now worth $31,000 and your $1,000 worth of Yahoo stock had ballooned to $46,000. Turning a $2,000 investment into

$77,000 is phenomenal on any time scale, but to do so in less than thirty months is unheard of. And the funny thing was, getting this sort of return wasn't exactly rocket science. In the twelve months of 1998, Yahoo stock returned 584%, AOL 593% and Amazon 970%.[5] These were three of the best-known, most talked-about stocks of the mid-nineties, widely heralded as the vanguard of the new economy that the Internet was supposedly bringing into existence. They were hardly needles in the haystack.

In the last two years of the nineties, seemingly any random Internet stock pick began to feel like a sure-thing lottery ticket, and that is why we remember this period as the dot-com bubble. Internet stocks proved to be particularly susceptible to speculation for a couple of reasons. Dot-com companies were young. They were going public sometimes only months after their creation. When they showed any sign of growth, their stock prices took off because it seemed to validate the notion that there was only more growth ahead. And it was that limitless promise that led to the second unique feature of Internet stocks: the profits didn't seem to matter. Valuations weren't tied to things like, you know, income. They were tied to potential fortunes to be made, somewhere in the future. New metrics like counting "eyeballs" and "mind share" were used to show companies were growing, even if that growth couldn't be measured in dollars and cents. Heck, sometimes a dot-com stock would increase in value even after it announced *losses*! Investors might take that as a sign the company was "wisely" plowing its money into strategies for growing at all cost.

Americans believed all this, because all the so-called experts were telling them it was true. This Time It's Different™ was a rallying cry of the time period. Magazines like *Wired* were promoting a glittering future where technology would soon be a panacea for all of mankind's ills. Books like Ray Kurzweil's *The Age of Spiritual Machines* promised that technology might help us transcend death itself. Bestsellers like *The Long Boom* and *Dow 36,000* made the argument that technological advances were enabling a structural shift that would kick the global economy into a new, higher gear, almost unfathomable to contemporary minds.

These arguments—that technology was changing the game and

that investment markets overall were being transformed—fused until they were almost one and the same, a self-reinforcing battle cry. All of this whipping up of idealistic hysteria found a willing accomplice in the financial press. On television especially, the gyrations and permutations of the boom were given literal play-by-play treatment by the channel that made its reputation during the late nineties. Early in the decade, CNBC had been an unprofitable, poorly watched channel on deep cable, the dorky, boring relation to CNN. But in late 1993, Roger Ailes took over the channel and transformed it. Taking his cue from the way that ESPN covered sports, especially with its *SportsCenter* franchise, Ailes began populating CNBC with winning personalities who covered the stock market the way a sports anchor might cover a bowl game. All through the day, a parade of talking heads from Wall Street came on to analyze fluctuations in the market. Today, we're used to cable news being a daylong parade of talking heads debating topics in *Brady Bunch*–style boxes. But before Ailes took this format to Fox News and it became standard operating procedure on cable news everywhere, the free-for-all gabfest format found its first success on CNBC.

By the turn of the century, CNBC had become the background noise for a particular American moment, the default channel of the bubble era. It was "an authentic cultural phenomenon," as *Fast Company* magazine described it, "broadcast to nursing homes, yuppie gyms, dorm rooms, hotel lobbies, pilot ready rooms, and restaurants" so that Americans could get a quick update on their favorite stock or the hot new IPO that was hitting the market. People at the time felt that CNBC was the most visible aspect of an overall democratization of investing that was taking place. "Why can't Joe Smith who works at a deli have the same information as Joe Smith who works at an investment bank?" said CNBC's Maria Bartiromo when asked to define her role to everyday investors. "That's why it's a bull market. It's not a professional's game anymore."[6] Years later, Maggie Mahar would concur. "It was in the last five years of the 90s that you saw the individual investor really take over," says Mahar. "They were really leading the market. They were doing a lot of the buying."[7] Indeed, the numbers bear this out. In a 2002 study, 40% of investors with financial assets of $25,000 to $99,000 reported making their first-ever stock purchase after January 1996. They were doing a lot of the buying because of the new online trading platforms that had proliferated, like E*TRADE, Ameritrade, Firstrade, Schwab, and more. By late 1999, the number of online brokerage firms was nearing 150, and normal Americans were making half-a-million online trades every day.[8] By 1999, nearly 40% of retail security trades were being done online.[9]

If Joe Smith saw a stock like Lycos profiled on CNBC, he could jump online and place an order for Lycos stock within minutes. There was no longer any middleman to talk him out of it. And if Mr. Smith wanted to spend his days discussing the relative merits and future prospects of Lycos, he could do so on message boards at sites like Yahoo Finance that had many thousands of forums devoted to discussing individual stocks. Often, the readership of these boards would break down between bulls and bears, or longs and shorts. Today, we are all familiar with the Roman Colosseum–like combat that goes on in the comments section of an average blog post, or the pages of a site like Reddit, but it was in the late nineties that average Americans became familiar with Internet conventions—such as flame wars and trolls—thanks to the bull versus bear debates on stock market–focused pages of a site like the Motley Fool.

■

IN DECEMBER 1998, a thirty-three-year-old stock market analyst by the name of Henry Blodget was working for the investment bank CIBC Oppenheimer.[10] Oppenheimer was not a particularly prominent player on Wall Street, and Blodget was not a particularly important analyst; he had basically lucked into the job less than three years previously, because banks were desperate to find someone "young" who understood this new Internet thing. Two months earlier, Blodget had published his first analyst report on Amazon.com. He had recommended buying the stock, setting a one-year price target of $150 a share. It was a good call. At the time of Blodget's first recommendation, Amazon was trading at $80 a share; it had subsequently exploded to $240. The Oppenheimer sales team wanted a fresh recommendation to take to their clients for the new year. At their behest, Blodget dutifully calculated that a 70% rise over the course of the next year might make sense, based on Amazon's recent sales growth. He put a new price target on the stock: $400 a share, writing, "Amazon's valuation is clearly more art than science, and we believe that the stock will continue to be driven higher in large part by the company's astounding revenue momentum."[11]

A far more experienced analyst covering Amazon at the time was Jonathan Cohen. Cohen worked at a more prominent firm, Merrill Lynch, and unlike Blodget, Cohen's analysis was widely followed. A few months previously, Cohen had actually downgraded his recommendation of Amazon to "reduce," saying the stock was too expensive. More precisely, Cohen would later, famously, call Amazon "probably the single most expensive piece of equity ever, not just for Internet stocks but for any stock in the history of modern equity markets."[12] Cohen's price target for Amazon was $50. So, Henry Blodget was going out on a limb by making such a wildly divergent call from the more experienced Cohen's. When Blodget circulated his numbers internally, "One of my bosses stopped by my office and sort of raised his eyebrows—'$400 a share?' " Blodget would remember later. The next day, when the call went public, "My phone lit up like a Christmas tree. I thought, 'Oh, no, I blew it.' "[13]

Far from blowing it, the Amazon call made Blodget's career. Blodget made his famous forecast of Amazon's $400 a share on December 16, 1998. The stock closed up 20% that day alone, in no small part thanks to news of Blodget's recommendation. By January 6, not even a month later, Amazon's stock blew past Blodget's $400 target. Almost overnight, Blodget became a regular on CNBC. He began to be routinely quoted and profiled in almost every newspaper and financial magazine in the country. A month later, when Jonathan Cohen left Merrill Lynch, Blodget took over Cohen's analyst chair at the more prestigious firm. By 2001, Blodget would be paid a rumored $12 million a year for his stock analysis.[14]

The experience of Jonathan Cohen was not unique on Wall Street. Hedge fund managers, mutual fund managers, stock analysts, even financial reporters learned and internalized a sharp lesson in the late nineties: People didn't want to hear negativity. For everyone involved, it was far more helpful to your career if you joined the hosanna chorus talking up the prospects of the soaring market. Fund managers who did not fill their holdings with technology stocks saw their returns trail those of their peers and even the market indexes. "You either participate in this mania, or you go out of business," Roger McNamee, one of the most famous technology investors of the era, told *Fortune* in June of 1999. "It's a matter of self-preservation."[15] One by one, bearish stock market analysts who for years had been saying the bull market was too good to last threw in the towel and got with the program.[16] Now one of the most famous technology stock boosters, Blodget joined a pantheon of Wall Street soothsayers who were almost ubiquitous in the late 1990s, analysts like Ralph Acampora, Jack Grubman, and especially Mary Meeker and Abby Joseph Cohen. Their slightest utterance could move markets, and they were all fully committed bulls, staking their reputations on the growth prospects of Internet companies.

Economists of all stripes were looking for a justification, a rationale, anything that could explain the boom times that they felt certain they were living in. Most just instinctively credited information technology. After all, everything was getting connected! The world was shrinking! Computers were everywhere! Surely that meant that things were functioning better, more efficiently, more profitably. The only problem was, none of this seemed to show up in any of the official numbers. Economic output is easy to measure when you can count widgets coming off an assembly line. But when your "economic revolution" is built around thoughts and ideas, and the speedy new ways you're connecting them all together, how do you quantify the value of those innovations? ATMs might mean fewer bank tellers had jobs; but think of the time saved by millions of consumers! How did one measure that? "More and more, value is produced not by real assets like factories and capital, but rather by people thinking and working together," *Fortune* opined in 1999. And yet, "while it seems obvious that computers have to have boosted productivity, proving that they have has been impossible."[17]

Many people came to believe that the proof might just *be* the soaring stock market. According to this line of thinking, stocks (and tech stocks especially) were rising because investors were rationally pricing in the vast improvements and profits that technology was making possible. Stock markets are a forward-leaning indicator of economic trends, and so perhaps the market itself was revealing the profits and efficiencies that would show up in official figures sometime down the road.

This rationale went all the way to the top. When Chairman of the Federal Reserve Alan Greenspan couldn't find the increases in productivity that he felt must be behind the run-up in stock prices, he commissioned Fed researchers to dig deeper into their statistical data in order to prove that productivity was, in fact, growing faster than government numbers showed. "Greenspan condoned the bubble—and then concocted a theory as for why it was rational," quips Maggie Mahar.[18]

Greenspan had begun the dot-com era skeptical of the stock market's euphoria. In December 1996, the Fed chairman gave a speech to a conservative think tank where a throwaway line ("But how do we know when irrational exuberance has unduly escalated asset values?") briefly caused markets to seize up.[19] "Irrational exuberance" would, somewhat ironically, become a cultural slogan of the dot-com era. But as the nineties wore on, Greenspan—if he did not exactly repudiate the phrase—gave every indication to the markets that he was no longer much worried about speculative excess. In January 1999, a senator asked Greenspan how much of the run-up in stocks was "based on fundamentals, and how much is based on hype?" The chairman answered: "You wouldn't get 'hype' working if there weren't something fundamentally, potentially sound under it." In the nearly two years after the "irrational exuberance" speech, the Federal Reserve raised interest rates only once, and, in fact, cut rates several times in response to the various mid-nineties "crises" few now remember, like the so-called Asian Flu of July 1997.[20] So, from late 1996 until late 1998—just the time when the dot-com bubble was inflating—the Fed was, to borrow from Wall Street lingo, extremely "accommodating" to the stock market.

Many people, then and now, feel that Greenspan, at the very least, enabled the dot-com speculative stock market bubble. At the time, American investors came to believe very strongly that Greenspan wanted them to be rich, and if anything went wrong, Uncle Alan would put his finger on the scales and make things right. During the run-up to the 2000 election, presidential candidate John McCain vowed: "And by the way, I would not only reappoint Alan Greenspan—if he would happen to die, God forbid—I would do like they did in the movie *Weekend at Bernie's*. I would prop him up and put a pair of dark glasses on him."[21]

■

IN THE WORDS of James Grant, editor of *Grant's Interest Rate Observer*, writing in 1996, "The stock market is not the kind of game in which one party loses what another party wins. It is the kind of game in which, over certain periods of time, nearly everyone may win, or nearly everyone may lose."[22] By the late '90s, everyone involved in the stock market seemed to be winning. And the coming of the dot-com stocks only seemed to extend this winning streak. Nobody had any vested interest in questioning the madness, least of all the media. As early as 1997, an estimated 30% of national newspaper ad revenues came from the financial services industry.[23] By 1999, ad rates on cable television were up 21% year-over-year and 16% on network television, thanks to an estimated $1.9 billion that young dot-com companies would spend to promote themselves.[24]

Most important, all those baby boomers, all those CNBC addicts, all those everyday Americans who were invested in the stock market—they were making money too. If they were invested in the right Internet stocks, they were making a *lot* of money. *Fortune* estimated that Internet fever was adding $301 billion to the U.S. economy by 1998, and another study estimated that 37% of all new jobs being created were thanks to the Internet.[25]

All told, approximately 50,000 companies would be founded between 1996 and 2000 aiming to commercialize the Internet, backed by more than $256 billion in venture capital.[26] But if the dot-com bubble is remembered mainly for the initial public offerings of stock that made all the headlines, it's important to remember that the actual dot-com mania, as measured by high-profile Internet IPOs coming to market, happened in a relatively brief window of time. In 1995, 7 stocks IPOed that could be termed "Internet companies." In 1996, there were 27. In 1997, the first of the real "dot-coms" came to market, totaling 19. In 1998, there were 29. But in 1999, there were 249 Internet IPOs. And those were just the Internet companies that debuted on the stock market. There were untold others that got acquired or went nowhere.

It was perhaps inevitable that, toward the tail end of the bubble, there were a lot of young Internet companies being founded that had questionable business plans at best. Some of the companies were so flimsy as to be just short of outright fraud. Investors (both venture capitalists and the public at large) no longer had any interest in discerning true value; any company with a .com at the end of its name might be the next billion-dollar winner. "You've got stocks selling at absolutely unbelievable multiples of earnings and revenues," the eternally skeptical old-school money manager Barton Biggs said as early as 1996. "You've got companies going public that don't even have earnings. You've got people setting up Internet pages to reinforce each other's convictions in these wildly speculative stocks."[27] By the end of the decade, such Chicken Little cries seemed quaint. If Americans—especially the everyday Americans who were in no way financial professionals, but were suddenly driving the market—were demanding to invest in Internet companies, Silicon Valley and Wall Street were more than happy to supply the demand. And with every new company that enjoyed a 100% first day "pop" on the markets, the increasingly isolated voices that were urging caution seemed all the more discredited. A well-respected, longtime stock market insider weighed in at the tail end of 1998, saying, "It defies my imagination that so many people with so little sophistication are speculating on these stocks."

The man speaking these words was Bernie Madoff.[28]

9

IRRATIONAL EXUBERANCE

The Dot-com Bubble

If you were looking for a single company that exemplified the dot-com era, you could do worse than Priceline.com.

Priceline was founded by Jay Walker, a forty-two-year-old entrepreneur with a clever solution to a real problem: every day, 500,000 airline seats were going unsold.[1] Priceline would offer these vacant seats to online customers who could name the price they were willing to pay to fill them. Consumers would (theoretically, at least) get cheaper flights; airlines would be able to sell excess inventory; inefficiencies would be ironed out of the market; and Priceline would take a cut for facilitating the whole process: your garden-variety win-win-win-win that only the Internet could make happen.

Launching in April of 1998, Priceline was a dot-com "overnight success," growing from 50 employees to more than 300 and selling more than 100,000 airline tickets in its first seven months of business. By the end of 1999, it was selling more than 1,000 tickets a day.[2] Believing in Amazon's Get Big Fast business strategy, Priceline attempted to expand into hotel bookings, car rentals, home mortgages—seemingly every market with excess inventory that a consumer might want to name a lowball price for. On the strength of this idea, Priceline was able to raise $100 million in working capital. Airline tickets were just the proof of concept. Walker's intention was to take this idea to every applicable market. "Priceline is just the beginning," he told the *Industry Standard*.[3]

Walker intended to get to ubiquity the way Yahoo had done: by building a brand through relentless marketing. In its first six months, the company spent more than $20 million in advertising, the keystone of which was clever radio and TV ads featuring *Star Trek*'s William Shatner.[4] The ads were reportedly scripted by Walker himself, and Shatner was compensated with 100,000 shares of stock instead of the originally offered $500,000 in cash ("Wasn't *that* a good move?" Shatner asked a *Fortune* writer in September 1999 when the shares were worth about $7.5 million).[5] All of this succeeded in placing Priceline fifth in Internet brand awareness by the end of 1998, behind only AOL, Yahoo, Netscape and Amazon.[6]

Forbes put Walker on its cover as a "New Age Edison." He told the *Industry Standard*: "The long-term legacy of Priceline [will depend on] whether or not we can successfully introduce the first new pricing system in probably 500 years."[7] In March 1999, Priceline went public at $16 a share, and on its first day of trading went up to $88 before settling at $69. This gave Priceline a market capitalization of $9.8 billion, the largest first-day valuation of an Internet company to that date.[8] After such a high-profile debut, few investors were concerned about the fact that in its first few quarters in business Priceline racked up losses of $142.5 million.[9] Or that it had to buy tickets on the open market—at cost—in order to fulfill the lowball bids its customers were placing, thereby losing, on average, $30 on every ticket it sold. Or that Priceline customers often ended up paying *more* at auction than they could have paid through a traditional travel agent.[10] Investors were more interested in grabbing a piece of a company that was going to change the future of business.

Because hey, by 1999, losing money was the mark of a successful dot-com. And few could lose money as prolifically or creatively as Priceline. The head of a rival travel website named CheapTickets complained that his company couldn't compete with Priceline's hype. "We've got a policy here at CheapTickets," founder Michael Hartley groused. "We need to make money. It hurts our valuation."[11]

Priceline's market valuation was doing just fine. At its highs,

Priceline had a market cap larger than any of the airlines it sold tickets for, and Walker's 49% personal stake in the company was worth as much as $9 billion.[12]

■

SO MANY OF THE COMPANIES that would embody what we think of when we remember the dot-coms shared some or all of Priceline's traits: a business plan that promised to "change the world"; a Get Big Fast strategy to reach ubiquity and corner a particular market; a tendency to sell products at a loss in order to gain that market share; a willingness to spend lavishly on branding and advertising to raise awareness; and, above all, a sky-high stock market valuation that was divorced from any sort of profitability or rationality.

The dot-coms that tend to have lingered in popular memory were the ecommerce companies, which, like Priceline, were targeting mainstream consumers. Amazon had effectively killed the category of books online, and so, hundreds of ecommerce companies were founded to become the "Amazon for X," where *X* was whatever flavor of retail one could imagine.

Children's toys were estimated to be a $22 billion annual market. (Yearly spend on toys per child? $350.)[13] And so, eToys took a crack at this segment. Of course, there were established players in the toy space already, especially Toys "R" Us and Wal-Mart. But then, Amazon had "Amazoned" Barnes & Noble, hadn't it? So, in a similar way, eToys cofounder Toby Lenk intended to establish an online beachhead before the incumbents could react. "We can out-Barbie and out-Lego the mass merchants out there," Lenk told a reporter.[14] By October 1998, eToys could crow about attracting as many as 750,000 visitors a month. Those were actually great traffic numbers for that time period, but, of course, not all of those visitors bought something. By December 1999, after more than two years in business, eToys could only boast lifetime revenues of $51 million. That was about as good as the combined yearly sales of seven Toys "R" Us real-world stores—and Toys "R" Us had nearly 1,500 stores worldwide.

No matter. eToys went public in May of 1999, selling 8,320,000 shares at $20 apiece. On the first day, the stock leapt to $85, before settling at $76, a 282% pop. eToys had a market capitalization of $7.6 billion, compared to Toys "R" Us's $5 billion. Toby Lenk's 7.36% share of the company was worth a cool $559 million.[15]

Entrepreneurs are always eager to grab a piece of the insane amount of money Americans spend on their furry friends ($23 billion in 1998; $60 billion as recently as 2015).[16] And so, as if out of central casting came four pet-centric entrants in the dot-com ecommerce sweepstakes: Pets.com, PetStore.com, Petopia.com and PetSmart.com. In February 1999, Pets.com was launched by an entrepreneur named Greg McLemore. If Get Big Fast was a matter of necessity for most dot-coms, it was especially so for Pets.com, as it was facing so many competitors. Pets.com enjoyed some powerful backers, including, coincidentally, Amazon.com, which took a 54% stake in the company.[17] The requisite IPO raised the company $82.5 million in February 2000, only a year after the company's founding. But, the devil was in the details. In the Pets.com IPO prospectus, the company stated that from the time of its inception through December 31, 1999, the company lost more than $61 million on sales of only $5.7 million. Why so much red ink? It didn't help that the cost of the $5.7 million in goods sold was $13.4 million. Pets.com was selling things for less than they cost! In fact, it was losing 57 cents on every dollar made in sales. It also didn't help that Pets.com's bestselling product—pet food—was a heavy, bulky item. Pets.com charged only $5 for shipping, even though the actual shipping cost of a 30-pound bag of kibble was reportedly twice that.[18] This was a not uncommon problem for the ecommerce players. A startup named Furniture.com raised $75 million only to learn a lesson that Ikea had known about for years: you can't exactly send a couch via UPS. "There were many cases when we would get an order for a $200 end table and then spend $300 to ship it," a former Furniture.com engineer would admit. "We never could figure it out."[19]

Okay—books, toys, pet food, furniture? What was left? How about one of the biggest retail markets imaginable? The total U.S. market for groceries, drugstore merchandise and prepared meals was over $650 billion in 1998 and by the end of the nineties, Americans were spending, on average, $5,000 a year on groceries, or 10% of their income.[20] Hoping to capture this spending by bringing it online were startups like Peapod, MyWebGrocer, Streamline and, especially, Webvan.

Webvan was the brainchild of a man who had already seen his previous business "Amazoned." Louis Borders was the founder of the bookstore chain Borders Group, Inc., and he was determined to do to grocery retailing what Amazon had done to book retailing. Borders knew that for every $100 in grocery store sales, $12 was eaten up by the cost of simply running the grocery store. In a famously low-margin business (for every $100 in sales, the typical grocery store sees only $2 or less in profit), eliminating a big cost center like that could be transformative. "Intuitively, I knew I'd have a great financial model if I could eliminate store costs," Borders told *Businessweek*.[21] And that was the promise of ecommerce, right?

Borders convinced Goldman Sachs, Benchmark Capital, SoftBank, and Sequoia Capital to invest a total of about $400 million in four rounds of venture financing, one of the largest capital raises of the dot-com era.[22] To test its concept, Webvan built a 330,000-square-foot warehouse in Oakland, California, to serve customers in the San Francisco Bay Area. The company also spent three years and hired eighty software engineers to design the inventory management, delivery and logistics systems required to make the operation function.[23] The idea was that once San Francisco proved the market, Webvan would expand to other cities and regions, building similar distribution centers, to the tune of $35 million per facility. Webvan promised that each distribution node would serve the equivalent customer base of eighteen conventional supermarkets, but with less than half the labor costs and double the selection of items.[24]

Launching in June 1999, Webvan began by offering prices it claimed were 5% lower than conventional grocery stores.[25] In order to entice customers, it often waived the delivery fee that was crucial to covering costs. In essence, Webvan tried selling groceries at a loss in order to achieve scale. But that was standard practice at this point, and in no way prevented Webvan from enjoying a typically buoyant IPO. When it went public in the fall of 1999, the company had recorded only $4 million in revenue in its entire existence. Nonetheless, the stock went out at $15 and rose to $34 before ending the day at $25. Webvan had an $8 billion valuation.[26] One executive from the competing grocery chain Safeway, which had been in operation for nearly a century and had hundreds of locations, complained: "They have the sales of two of our stores and one-fourth of our market cap."[27]

Webvan stated that if approximately 1% of Bay Area households, about 120,000 families, used its service on a regular basis, it would be profitable. The problem ended up being that even though about 6.5% of Bay Area households tried Webvan at least once, only half that number ever placed a second order, and even fewer became weekly or even monthly customers.[28] The distribution centers needed to operate at 50% capacity in order to cover costs. But by the first quarter of 2000, the Oakland warehouse was operating at only 35% capacity and reported a $38.7 million loss.[29] Webvan nonetheless ignored these hiccups and barreled ahead, opening additional warehouses serving Atlanta, Chicago and Sacramento, where the losses only widened. None of the warehouses reached an order volume that allowed them to break even, and by the spring of 2001, the company was losing $100 million a quarter.[30]

Of course, even this unfolding high-profile disaster didn't stop other entrepreneurs from chasing the same dream. On the East Coast, two companies, Kozmo.com and UrbanFetch, took instant gratification a step further: both promised same-day delivery. But the question was, could anyone make money doing that? That pint of Ben & Jerry's a customer ordered on a rainy afternoon? Kozmo would send it to them for less than it would cost to buy at the local bodega across the street. And Kozmo still had to pay the army of bike couriers who made the delivery. It was retail without the overhead of real estate, sure, but what about the costs of warehousing, of labor, of the website and logistical back-end systems? Neither Kozmo nor UrbanFetch were much worried about this. Ubiquity came first. Profits later.

Again, no one was focused on inconvenient details like the costs of doing business or profit margins. Investors, entrepreneurs, venture capitalists and Wall Street tended to prefer numbers like those from an OECD report in 1999, which assured everyone that by 2005, online commerce would be a $1 trillion market, representing 15% of overall retail sales. So, hurry up! Stake your claim! There was nothing but growth ahead, so if you locked consumers in with low prices now, you could always raise prices later, once you had killed your category.

For a couple of years there, it seemed like everyone was begging us to buy cheap stuff, subsidized largely by generous, unseen piles of venture capital money. Ironically enough, far from engendering customer loyalty, consumers tended to treat the dot-coms as a fly-by-night bonanza, taking the deals when they presented themselves, but often not repeating the experience. "They all e-mail me specials," one New York pet owner told *Businessweek* of her experience with the pet dot-coms. "I order from whoever has the special. Sometimes, it's even free."[31]

Pets.com was losing money on every dog leash it shipped. But if you looked at the company's bottom line at the time of the IPO, the biggest expenses, at $42.5 million—a whopping 76% of total operating costs—were for marketing and sales. Advertising. And that's why we remember Pets.com, if, indeed, we remember it at all. Priceline might have had William Shatner, but Pets.com had the sock puppet.

Soon after launching, Pets.com hired the ad agency TBWA\Chiat\Day to produce a reported $20 million initial ad campaign.[32] TBWA had recently produced a series of ads for Taco Bell featuring a talking chihuahua, and, perhaps taking a page from that campaign's success, the ad men proposed a talking dog–like sock puppet that would commiserate with real-life pets in a series of commercials (tagline: "Pets.com. Because pets can't drive."). The puppet was voiced by the comedian Michael Ian Black, but was deliberately nameless, "so consumers would always have to say 'Pets.com' when referring to it."[33] Soon, the puppet was airing in radio and television spots nationwide. Pets.com paid nearly $2 million for an ad on Super Bowl XXXIV and the puppet became a float in the 73rd Annual

Macy's Thanksgiving Day Parade.[34] After appearances everywhere from *Live with Regis and Kathie Lee* and *Good Morning America*, to "interviews" in the pages of *People* and *Entertainment Weekly*, Pets.com began to license the puppet as a popular toy for children.

In a single quarter, Pets.com reportedly spent $17 million promoting the sock-pooch. Was it worth it? Well, not when you consider that in that same quarter it had only $8.8 million in total revenue.[35] By October of 1999, Pets.com was third in the race for website visitors among the pet competition, attracting only 551,000 unique visitors (behind leader Petsmart.com's 1.1 million), and it was paying $158 for every new customer it acquired.[36]

■

DOT-COM COMPANIES FELT they had to spend in order to brand themselves like Yahoo had done. They felt they had to be first to their particular market in order to lock in customer loyalty, just as Amazon had done. They spent because they felt they had to be the first in their category to IPO, like eToys had. Spending big on marketing could help you get that IPO. And then, after the IPO happened, it could help keep your stock price high. "You could reasonably argue that every additional $1 of revenue this quarter might increase your market capitalization by $300 next quarter," PetStore.com's Josh Newman said.[37] Higher stock price, higher market cap: more money, both tangible and on paper. Spending, spending, spending became a vicious cycle that artificially turbocharged everything in the dot-com era. It became a joke that the very dot-coms that started out promising this grand vision of a more efficient way of doing business were—almost to a company—unprofitable. It's entirely possible that a lot of them could have focused on the very real efficiencies that selling online made possible, and thereby slowly grow sustainable businesses. But that was not the name of the game in the late nineties. The name of the game was Get Big Fast.

The venture capitalists who backed these companies were aiming for supernova IPOs, because that's when they got paid. Any IPO meant an "exit" for venture investors. Those incredible first-day "pops" that dot-com stocks experienced when IPOing? That was the early money cashing out, selling their shares to the investing public, who would now be holding the bag, waiting to see if that fancy new business model would ever work out. The dot-com bubble was a fantasy period when a lot of VCs actually *didn't care* if a business model made sense, because it didn't need to. "We're in an environment where the company doesn't have to be successful for us to make money," a venture capitalist at Benchmark admitted when mulling over a pre-IPO investment in Priceline.[38]

It became imperative to keep the pipeline of new companies—and therefore, new IPOs—coming. Fortunately enough, the bubble era engendered a sort of fever for entrepreneurship that probably hadn't existed in this country since before the Great Depression (the Roaring Twenties, the age of the tinkerer-developers of the automobile, the telephone, the radio, the airplane). By the spring of 1999, one in twelve Americans surveyed said that they were in some stage of founding a business.[39] If so many of these new entrepreneurs were chasing the fortunes that dot-coms seemed to be minting every day, who could blame them? In 1994, the venture capital firm Draper Fisher Jurvetson received 376 business plan proposals. By 1995, the year of Netscape's IPO, that number had reached 1,075. By 1999, there were more than 12,000 business plans to sift through.[40] The supply of entrepreneurs was more than met by eager venture capitalists who were all but begging the new companies to take their money. In 1998 alone, 139 new venture funds were created, with more than $17.3 billion in new capital to invest with, an increase of 47.5% over the previous year.[41] "It was absurdly easy," a young Harvard Business School graduate said of the fundraising process during the dot-com era. "You would walk into offices in New York and people would immediately offer money to you if they thought you looked smart. We didn't have any data on the market; we didn't have a product demo; we didn't have anything. We had a business plan, but that was it."[42]

Venture capitalists know that they have to kiss a lot of frogs before they find a prince, but the dot-com era was a uniquely good time for VCs, because the willingness to take companies public under any circumstances—profitability be damned—meant that VCs weren't punished for being indiscriminately promiscuous. Even the ugliest frogs could be winners. The average yearly return for venture funds that focused on early-stage startups was 25% by 1998, and plenty of the top-tier funds were earning well in excess of 100% or 200% yearly on invested capital.[43] VC is a game of blockbusters; one home-run investment like an eBay, returning 100,000%, can make up for a lot of losers. And even then, what did it matter if you backed a loser when you could take it public and cash out one way or another in less than nine months?

Over the course of the entire 1980s, IPOs rose on average 6% on their first day of trading and there had only been seven IPOs that had doubled.[44] In the first quarter of 1999, Internet IPOs gained an *average* of 158% on their first day.[45] In the first quarter of 2000, technology companies were going public and doubling, just about every other day.[46] Several companies we've mentioned in earlier chapters benefited from this IPO mania. MarketWatch went public on January 15, 1999, and enjoyed a 473.5% first-day pop; iVillage, on March 19, 1999, 233.9% pop; Broadcast.com, on July 17, 1998, 248.6%. And the IPO madness didn't mean the takeover madness ended. On the contrary, it intensified. Broadcast.com and GeoCities had enjoyed successful IPOs (a 119.5% pop for GeoCities), but the founders of both companies eventually succumbed to takeover offers they couldn't refuse. In January 1999, Yahoo paid $3.6 billion to acquire GeoCities. At the time, GeoCities was generating only $7.5 million a quarter in revenues and had no profits.[47] But Yahoo followed this up by purchasing Broadcast.com in April, in a deal then valued at $6.1 billion, or 474% more than the value of the company on the day of its IPO. Why did Yahoo do these deals? For traffic. For eyeballs. At the time, GeoCities had 19 million unique monthly visitors, making it the third-most-trafficked site in the world behind AOL and Yahoo itself. In the case of Broadcast.com, Yahoo was purchasing the most mature play in the world of streaming media. The portal was bulking up in anticipation of doing battle with AOL to become the premier media company of the twenty-first century.

Of course, Yahoo could afford it. With all of the advertising money flowing in from other dot-coms, and the portalization efforts paying off to the tune of traffic numbers approaching 100 million unique visitors per month, Yahoo's market cap surpassed $120 billion at its peak around the turn of the millennium.[48] Its price-to-earnings ratio got as high as 1,900.[49] It had plenty of money to throw around. Woe be to the other portal sites that had to keep up!

Perhaps the most incredible deal of the time was Excite@Home's acquisition of Blue Mountain Arts for $740 million dollars in cash and stock. Excite@Home was a company formed when the broadband ISP @Home merged with the search portal Excite.com. Blue Mountain Arts operated the website Bluemountain.com, where users could send each other electronic greeting cards by email. That's right. Bluemountain did nothing but send Grandma electronic "get-well-soon" greetings. But Bluemountain.com was getting 9 million unique users a month to do this, and at the time, traffic was the sine qua non for a Yahoo-chasing portal player like the Excite half of Excite@Home.[50] As the *New York Times* noted in its article announcing the deal, Excite@Home "predicted that the acquisition would increase its audience by 40%, to encompass approximately 34% of Internet traffic."[51] So, Excite@Home was willing to pay $82 per user to attract additional eyeballs to its network of properties and try to keep pace in the portal race.

The merest association with the word "Internet" could suddenly make a company seem more valuable, as when K-Tel, the "as seen on TV" music retailer of such music series as Hooked on Classics, announced that it was launching a website to market its CDs over the Internet. K-Tel stock went from $3.31 to $7.46 in a single day. Less than a month later, it was trading at $33.93.[52] Nothing fundamental had changed in K-Tel's business. It had merely launched a website. A similar thing happened with Active Apparel, owner of the boxing and activewear brand Everlast. When it announced an ecommerce website, its stock exploded by more than 1,000% in the following two trading days.[53]

In the midst of this sort of frenzy, there was space for plenty of dubious companies to receive funding. iHarvest.com was able to raise $6.9 million to create a tool for web surfers to save copies of web pages for later offline browsing. This, despite the fact that almost all browsers already had bookmark buttons.[54] Iam.com raised $48 million to host the headshots and portfolios of aspiring actors and models.[55] Officeclick.com raised $35 million to create a community site for secretaries and administrative professionals. Other companies continued to take stabs at reinventing ecommerce. Mercata.com raised $89 million to create a group-buying marketplace where thousands of people would buy items in bulk in order to get better pricing. One day after its IPO was canceled, the company declared bankruptcy.[56]

If Mercata sounds like an eerily similar idea to later social-buying companies like Groupon, that's not exactly unusual. Plenty of dot-com startups were founded around concepts that were quite possibly good ideas but were just a bit too early for the time. eCircles.com pioneered online photo albums, and Myspace.com and Desktop.com rented what were essentially virtual hard drives—what we now call cloud storage. After going bankrupt, the Myspace.com domain would later be put to use by another startup we'll discuss shortly.

A lot of companies were nothing more than IPO plays. And in the worst instances, some of the bubble companies were platforms for outright fraud. Pixelon was a company that raised $35 million in venture financing, promising to develop "full-screen, TV-quality video and audio streaming technology" in an era of dial-up modems.[57] It promptly turned around and blew $16 million of that on a company launch party at the MGM Grand in Las Vegas that featured performances by KISS, the Dixie Chicks, Sugar Ray, and a reunion concert by the Who. It was later revealed that Pixelon founder Michael Fenne was *really* a man named Paul Stanley (no relation to the guitarist from KISS) who was wanted by the State of Virginia on stock fraud charges. Pixelon never released a product before it was eventually forced into bankruptcy.

The parties, the hype, the headlines, it was all part of the milieu. In any fad or bubble, eventually the scenesters show up. And when the pretty people arrive, that's usually a sign that a bubble is at its height. This was especially true in the media capital of the world, New York. And if one company exemplified hype-as-a-business-plan, it was Pseudo.com. Pseudo was the brainchild of Joshua Harris, a technology early adopter who had previously founded the tech research company Jupiter Communications. Pseudo's stated goal was quite simple: to bring television online. To this end, Pseudo invested in studios and creative talent to produce dozens of different "shows"—about 240 hours of original programming a month—that it broadcast over the web from its SoHo headquarters.[58]

The shows that Pseudo produced ran the gamut of subjects, from sports to video games to music to talk shows. Pseudo combined video with online chatrooms to create programming that was self-consciously interactive. The on-air talent mixed freely with the viewers who lurked in the chatrooms and often impacted what was happening on air, in real time. Like a public access channel on hallucinogens, Pseudo claimed it was establishing an entirely new medium that would be like the second coming of television—but two-way and interactive.

If producing television for the twenty-first century was the stated goal of Pseudo, the delivery method seemed to be a 24/7, never-ending party. Harris and Pseudo became, briefly, ground zero for the New York City art scene, and Pseudo's regular events and parties put Pixelon's Las Vegas bash to shame by rivaling the artiness and excess of Warhol's Factory ("I think I'll be bigger actually [than Andy Warhol]," Harris said).[59] The Pseudo soirees featured DJs, poetry and art, but also computers and video games. "I remember that some exhibitionistic fat guy with a really tiny penis started taking a shower while dinner was going on," said a gossip writer the *New York Post* dispatched to report on one Pseudo event. "The food was quite good, but I couldn't really enjoy it because some half-naked people who seemed to think they were very important kept dancing on the table."[60] These fin de siècle bacchanals were all funded by Harris and the more than $25 million that he was able to raise from the likes of Intel and the Tribune Company, ostensibly to turn Pseudo into a broadcaster for the twenty-first century.[61]

Silicon Valley was comfortable celebrating the dot-com companies with unquestioning adulation. After all, the Valley's whole industry is predicated on churning out the new. But New York was especially susceptible to dot-com envy, and it was there that the backlash against the bubble first began to take root. Journalists and old-media types began to look jealously at these kids, with their raves and their computers and their stock options that made them (on paper at least) worth millions of dollars for—what, exactly? Or, they could look at peers like iVillage founders Nancy Evans and Candice Carpenter Olsen, both of whom had come from publishing but had crossed the divide into digital moguldom, and were now pictured smoking enormous cigars in the pages of magazines after celebrating their record-breaking IPO.

When former Surgeon General of the United States C. Everett Koop became the eponymous public face of Drkoop.com, it must have felt like a thumb in the eye to any media celebrity who hadn't been smart enough to jump on the dot-com bandwagon sooner. Drkoop.com was nothing more than a general-interest health portal with a celebrity figurehead. Its traffic numbers were nothing special, and of course the site didn't make any money. Nonetheless, Drkoop.com enjoyed a nearly 100% first-day IPO pop and raised $85 million from investors despite reporting lifetime revenue totaling only $43,000.[62] Following this lead, veteran news anchor Lou Dobbs shocked the media world in June of 1999 by leaving his decades-long stint at CNN to launch Space.com, a space-focused portal financed by VC firms Greylock and Venrock Associates.[63] "I think most of the people here would be very insulted if somebody said the reason they are here is because of the potential of an IPO," Dobbs said of the company he quickly staffed up to about thirty employees. "I'm not saying that's not part of the equation, but it sure as hell isn't the primary reason," Dobbs was quick to add.[64]

By 1999, the faces in the annual list of the "Silicon Alley 100" included the usual suspects like Kevin O'Connor and Dwight Merriman of DoubleClick and Craig Kanarick and Jeff Dachis of RazorFish, but also Sam Donaldson of ABC News, who, late in 1999, launched a fifteen-minute, thrice-weekly, web-only video news show.[65] The Silicon Alley 100 was the yearly status list of the magazine *Silicon Alley Reporter*, launched by the New York tech gadfly Jason Calacanis to cover the New York tech scene with a slavish vigor that was intended to rival the way *Vanity Fair* covered Hollywood. Calacanis's magazine came to be seen as the calling card of what appeared to be a new media establishment, with Calacanis as the new media maestro.

At the end of 1999, in its final issue of the twentieth century, *Time* seemed to make the supremacy of the dot-coms official when it named Amazon's Jeff Bezos as its Person of the Year. At age thirty-five, he was the fourth-youngest person to receive this accolade, after Charles Lindbergh, Queen Elizabeth II and Martin Luther King Jr.[66] James Kelly, *Time*'s deputy managing editor, wrote that Bezos had been selected because "he has helped guarantee that the world of buying and selling will never be the same."[67] When he was asked if it truly was his intention that Amazon would one day to be able to sell anything, any item, Bezos responded: "Anything, with a capital A."[68]

■

BY OCTOBER 1999, the market cap of the 199 Internet stocks tracked by Morgan Stanley's Mary Meeker was a whopping $450 billion, about the same size as the gross domestic product of the Netherlands. But the total annual sales of these companies came to only about $21 billion. And their annual profits? What profits? The collective losses totaled $6.2 billion.[69] "People come in here all the time and say, 'The last thing I want to be is profitable,'" one investment banker bragged in June of 1999. "'Because then I wouldn't get the valuation of an Internet company.'"[70]

The continued craziness of the market, coupled with the increasing dubiousness of the companies and stocks that were going public, eventually pushed the bubble toward its end point. Over the second half of 1999, it wasn't a question of whether or not a bubble existed, it was a question of how big a bubble it was, and when it would pop. The entire nation seemed to be engaged in a "greater-fool" standoff. You bought stock or founded a company because you knew everyone else was doing the same. Most people knew the irrational exuberance was unsustainable, but no one wanted to be the first to admit it. After all, if you could squeeze your IPO out before the window closed, or if you could hold your Yahoo stock long enough for it to double one last time, then you could pick your moment to cash out, hopefully before everyone else got the same idea. In the meantime, you kept your own counsel and shook your head quietly as the last flood of dubious companies rushed the public markets.

Sensing this cynicism, the backlash among the New York media establishment began to creep onto Wall Street. *Barron's* came out with a widely read cover story analyzing the balance sheets of especially the ecommerce companies and warned that investor patience with continued losses was probably running out. This was coupled with distressing quarterly reports from some of the weaker dot-coms that sent their stocks downward. Even the big names began to come in for questioning. Another highly publicized *Barron's* cover story was titled "Amazon.bomb" and said, "Investors are beginning to realize that this storybook stock has problems."[71] If Amazon, the standard-bearer for the dot-coms, was in trouble, what did that mean for everyone else? A Lehman Brothers analyst named Ravi Suria began writing scathing reports questioning Amazon's very solvency as a going concern. Suria wrote that Amazon would likely run out of cash within four quarters "unless it manages to pull another financing rabbit out of its rather magical hat." The *New York Post* headlined, "Analyst Finally Tells the Truth About Dot-Coms." Around the time Jeff Bezos was feted as *Time*'s Man of the Year, Amazon's stock hit its all-time high, a split-adjusted $107 a share, and then slowly began to drop in price.[72] In February 2000, Wall Street was shocked when Amazon announced it had sold a $672 million convertible bond offering.[73] Why did Amazon need so much cash, unless it feared it was running out?

For the better part of two years, the dot-com mania had been fueled by the This Time It's Different™ mass faith that Americans had in the promise of the Internet. That sort of new-economy mumbo-jumbo worked for the dot-com companies—until it didn't. Get Big Fast and profits-someday were valid business strategies—until they weren't. The hundreds of new companies created in the dot-com era simply pushed credulity a bit too far, for a bit too long.

The flood of crap companies, especially those that came to market near the end of the bubble, could not be ignored forever. If the "good will out," as they say, then the opposite is true as well: the bad will out eventually, if given enough time.

One by one, the weakest of the dot-coms, those with the flimsiest business plans, or those that were the most blatant copycats of other flimsy ideas, began to underperform the market. Dot-coms ceased being sure stock market winners—at first in a trickle, and then all at once. Falling stock prices turned into stock market delistings and then became actual bankruptcies. Like any good game of musical chairs, when the music stopped, there simply weren't enough seats for everyone. As investors suddenly began to demand that companies show a profit for the first time, the collective response from the dot-coms was "What? You can't be serious!"

10

POP!

Netscape vs. Microsoft, AOL + Time Warner and the Nuclear Winter

Careful readers will notice that in all this talk about the dot-com frenzy, there hasn't been a single mention of the original dot-com company: Netscape. That's because even before the dot-com bubble was properly inflated to its greatest extent, Netscape had ceased to be an important player in events as they unfolded.

Netscape actually tallied impressive revenue growth in its first ten quarters as a public company.[1] It was, for a time, the fastest-growing software company in history, going from zero to half a billion in revenues in three years.[2] But that growth papered over the internal problems that later revealed that Netscape as a company was confused about its ultimate strategy. Netscape had IPOed as a software company: it developed web browser software that consumers and businesses ostensibly paid to use. At the time of its IPO in 1995, fully 90% of the company's revenues came as a result of its stand-alone Navigator web browser.[3]

But then came Microsoft and Internet Explorer. So, Netscape pivoted to service corporate customers with commerce servers and Intranet servers. By 1997, the percentage of Netscape's revenue generated by the stand-alone browser was below 20%.[4] The only problem with that state of affairs was that selling to corporations required a traditional, corporate-style salesforce. From 15 salespeople in 1995,

Netscape's sales army ballooned to almost 800 people by 1997, and sales and marketing costs ate up about 47% of revenue.[5] From the nimble and efficient "new-style" software company that Marc Andreessen and Jim Clark had told the press Netscape was destined to be, the company actually evolved into the very thing it had once ridiculed: a lumbering and inefficient "old-style" software and services firm.

"I absolutely thought we were a software company—we build software and put it in boxes, and we sell it," Marc Andreessen said in May 1998. "Oops. Wrong." The Netscape that had kicked off the Internet Era was now Netscape, the many-headed hydra, groping desperately for any business model it could find. "We've completely changed," Andreessen said.[6]

The irony was that the very company that had announced to the world that there were riches to be found on the Internet couldn't find a reliable way to make money on the Internet. After ten quarters of growth, Netscape's revenue suddenly dropped by 17% in the fourth quarter of 1997. In January of the next year, the company reported a quarterly loss of $88 million and laid off 300 of its 2,600 employees.[7] After reaching an all-time high in early 1996, by 1998, when the Yahoos and eBays of the world were entering the stratosphere, Netscape's stock was languishing below its IPO price.[8]

The question was, did Netscape stumble, or was it pushed? In October 1998, Microsoft's Internet Explorer passed Netscape's Navigator (later called Netscape Communicator) in browser market share.[9] Each new version of Internet Explorer released copied features that Navigator had pioneered, and then added features that Navigator didn't have. Microsoft usurped the browser market by giving away its browser for free. And more than free, there were instances where Microsoft was essentially paying valuable partners—Internet service providers, computer manufacturers—to favor IE over Navigator. Netscape, the smaller company by far, couldn't afford to give away its browser. The whole reason that Netscape tied itself in knots trying to reinvent its business model was that it knew it couldn't match Microsoft's deeper pockets when it came to competing in the stand-alone browser market.

In a last-ditch effort to shore up market share, Netscape released the source code to its browser on a website called Mozilla.org in January 1998. *The Economist* magazine said that this move was "the computer-industry equivalent of revealing the recipe for Coca-Cola."[10] This open-source browser project would later evolve into the Firefox web browser, which would, in the 2000s, eventually take the market-share crown back from Internet Explorer. But it did nothing for Netscape at the time. By February 1998, Netscape's stock was down by half from its IPO, and 88% off its all-time high.[11]

In a bit of asymmetrical warfare, Netscape had, very early on, turned to the federal government in an attempt to gain some sort of relief from Microsoft's predations. It sure as heck seemed to Netscape like Microsoft was leveraging its operating system monopoly to kill the market for web browsers. On August 12, 1996 (the same day that Microsoft shipped Internet Explorer version 3.0), Netscape sent a letter to the U.S. Department of Justice, claiming that Microsoft was wielding Windows 95 like a cudgel, preventing Netscape from doing deals with vendors and manufacturers that would allow the company to protect its place in the market. On October 20, 1997, the Department of Justice announced that it was investigating Microsoft for violation of a previous consent decree, and in May 1998 the attorneys general of twenty states joined the DOJ in filing antitrust suits against Microsoft.

The ensuing Microsoft trial was like a bonfire-of-the-vanities-style sideshow playing out in the background during the headier months of the dot-com bubble. Running from October 1998 to November 1999, the trial provided plenty of entertainment for those in the technology industry who both feared Microsoft and were jealous of it. The trial uncovered over 2 million emails, memos and other materials from within Microsoft, Netscape, and even other companies, such as AOL and Apple.[12] The government focused largely on proving that Microsoft had strong-armed companies into shunning Netscape, such as when AOL was induced into double-crossing Netscape over its default browser, and when computer manufacturers were cajoled into removing Netscape's Navigator as a preinstalled option.

The trial was intensely embarrassing to Microsoft executives, who, time and again, were contradicted by their own emails and previous statements. Not even Bill Gates was immune. The government played hours of a videotaped deposition from Gates, showing him sparring with the government's lead attorney, David Boies. Gates came off as dissembling, petulant, even petty. Like Bill Clinton's famous testimony dispute over what "the meaning of the word 'is' is," Gates argued over the characterization of basic words in his own emails. He denied remembering meetings and claimed to forget details about strategy—things that no person with a passing knowledge of the way Gates managed Microsoft could believe. Gates denied seeing Netscape as a serious competitive threat, in direct contradiction to previous public statements.

When the judge, Thomas Penfield Jackson, finally delivered his decision in the case, it was the verdict that Microsoft's enemies had been hoping for for years. Judge Jackson found Microsoft guilty of violating U.S. antitrust laws. Microsoft had "maintained its monopoly power by anticompetitive means and attempted to monopolize the Web browser market."[13] The suggested remedy: Microsoft should be broken up into two separate companies: one that developed and sold operating systems, and another that developed and sold applications like web browsers.

Of course, it never ended up that way. The case was appealed: the original verdict was rejected; and by the time the new Bush administration took office in 2001, there was little appetite for continuing what could be seen through a partisan lens as "antibusiness" litigation. Microsoft was never broken up, instead eventually agreeing to a Department of Justice settlement that required Microsoft to open its APIs and protocols, and generally play nice with competitors in the near future. Critics saw this as little more than a slap on the wrist.

With the benefit of twenty years, it's easy to look back on the Microsoft antitrust trial and even the whole Netscape/Microsoft web browser war as a bit of a tempest in a teapot. After all, we now know that Microsoft was about to enter something of a "lost decade" during which its influence on the industry would wane and the company would come to be seen by many as almost an irrelevant force as technology evolved. Indeed, Microsoft's diminished stature over the course of the 2000s would seem to validate one of the company's key claims during the trial itself: that the technology industry is so dynamic, so competitive, that no player, no matter how dominant in one market or at one point in time, can really be thought to be monopolistic. Because, in the blink of an eye, that entire market could change thanks to the arrival of new competitors or new technologies.

But from another perspective, it's worth wondering how much the flowering of the dot-com era was enabled by the fact that the most dominant, rapacious player in the industry was distracted while the new era was taking shape. The fact is, while Microsoft made plenty of moves during the dot-com era (MSN, Expedia, Hotmail, WebTV, just to name a few), it largely refrained from engaging in direct combat with the major dot-com players. More important, Microsoft never had the chance to absorb any of the cream of the new crop, as it had shown it was wont to do in earlier technology eras. Microsoft never attempted to acquire Amazon, say, though it certainly had the money to do so early on. And, crucially, once the dot-com bubble burst, Microsoft was in no position to swoop in and gobble up the wounded survivors because it feared angering the government again.

In short, it's easy to see, especially based on recollections that have come out from ex-Microsofties, that the antitrust trial hobbled Microsoft strategically, and maybe even creatively. "It had a big impact, and even a decade later it was still having an impact," Mary Jo Foley says of the antitrust trial. A journalist who followed Microsoft through the 1990s and 2000s, Foley argues that after the trial, no matter what product or feature it looked to develop, Microsoft had to think about legal issues first.[14] And so, one must consider to what degree Microsoft was distracted by the trial, allowing it to miss, say, the development of paid search as a dynamic new market, or the rise of social networks as an entirely new paradigm.

This is not to suggest that Microsoft was unsuccessful in the 1990s. On December 30, 1999, Microsoft reached its peak market cap of $618.9 billion.[15] This was due in large part to the fact that the 1990s was the decade that truly saw computing go mainstream. In 1990, there were around 9.5 million PCs sold in the United States. By the year 2000, that number had increased to 46 million annually.[16] Considering that around 90% of those machines were running Microsoft Windows and Office, no one was crying for Microsoft. In that same year, 2000, PC penetration in U.S. households passed 50% for the first time.[17] The web revolution helped normalize computing and Microsoft rode this wave as much as any dot-com. But the end result of the trial was that, going forward, Microsoft was merely another passenger; it was no longer steering the wave's direction.

■

IF MICROSOFT'S HEGEMONY over the tech industry was broken by the end of the decade, the only meaningful casualty of this structural earthquake was Netscape. The antitrust trial was, of course, not designed to save the fortunes of Netscape; the parties involved in the trial were the U.S. government and Microsoft. Netscape was just the star witness, the primary victim. And by the end of the trial, Netscape was not even an independent party anyway. On November 24, 1998, America Online announced it was purchasing Netscape for $4.2 billion in stock.

It was painful for many involved with Netscape that the pioneering web company couldn't overcome Microsoft's might. But the fact that Netscape ended up swallowed by AOL, the "training wheels for the Internet," seemed especially ignominious. "I mean, OK . . . Microsoft? A worthy opponent!" says original Netscape engineer Aleks Totic. "Did they fight fair? No, they did not. But . . . it's understandable. Now, being in a market where Netscape got sold to AOL? That was just depressing."[18] Most of the original Netscape team left the company rather than join AOL. Marc Andreessen briefly stayed on as AOL's chief technology officer, but he also left within a year, to form a new startup with fellow Netscape refugee Ben Horowitz. For his part, Netscape's other cofounder, Jim Clark, was already ensconced in his third billion-dollar startup, Healtheon (later, WebMD), which enjoyed one of those 292% first-day pops when it IPOed in February of 1999.

The Netscape folk might have looked down their noses at AOL (AOL never got much respect from true techies in Silicon Valley), but by 1998, their jaundiced view was not shared by Wall Street. AOL was how around 40% of U.S. users got online.[19] It was the most popular ISP in America—all the more so after it acquired its oldest rival, CompuServe, in September of 1997.[20] By the end of 1999, AOL would surpass 20 million subscribers.[21] At any given moment, especially in the evenings, as many as 1.1 million Americans were logged in to AOL.[22] And what so impressed Wall Street was that AOL was not only one of the few profitable Internet companies—by the height of the dot-com era, it had become a *really* profitable company. At the end of its fiscal year 1999, AOL could boast cash flow of $866 million dollars on revenue of $4.8 billion.[23] *Fortune* put Steve Case on its cover under the headline "Surprise! AOL Wins."[24] As the 1990s came to a close and Microsoft was distracted by its trial, it seemed to many in the industry that if there was any company that might be the heir to Microsoft's throne, it was AOL.

AOL had those 20 million Americans paying $21.95 a month to log in—a nice, steady stream of reliable revenue—but also had learned a new trick: advertising. By 1999, the company was generating $1 billion a year just in ecommerce and revenue deals—more than ESPN and ESPN2 combined.[25] Analysts were predicting that by 2003 AOL would generate more ad revenues than ABC or CBS.

No company took greater advantage of the bubble madness than AOL, by straight-up cannibalizing other dot-coms. You might remember that the Drkoop.com IPO raised $85 million for the health-information website. A month after its stock market debut, Drkoop turned around and basically spent all that money by agreeing to pay AOL $89 million over four years to provide health content to AOL users.[26] And that wasn't even the biggest deal AOL struck in those days. A long-distance phone provider named Tel-Save ponied up $100 million.[27] AOL skillfully played one competitor off another: Barnes & Noble paid $40 million to be the bookselling partner on

AOL's online service; Amazon paid $19 million just to be on the AOL.com portal; in the midst of fending off auction competition from Amazon, eBay ponied up $75 million for a four-year auctions exclusive. And Wall Street rewarded such tie-ups. Tel-Save's shares leapt from $13 to $19 after announcing its deal; Drkoop.com's deal announcement caused its stock to surge 56%.[28]

Locking down guaranteed traffic from AOL became a box that dot-coms had to check in order to begin the Get Big Fast sweepstakes. And as AOL realized the position of power it had over the dot-coms, the deal-making only got more aggressive. In 1998, the dot-com startup N2K tried to land a $6 million agreement for the privilege of being AOL's premier music retailer when, in the midst of negotiations, its executives let slip that they were in a hurry to close the deal ahead of N2K's forthcoming IPO. AOL promptly jacked the price of the deal up to $18 million, which represented more than ten times N2K's annual revenue.[29] N2K didn't even flinch. It paid the $18 million rather than risk a busted IPO.

AOL became so proficient at doing these deals, so rapacious, in fact, that it gained a reputation for aggressiveness that, until recently, only Microsoft had enjoyed. AOL's army of deal-makers were known internally as the company's "hunter-gatherers," because they descended on the dot-coms like predators and made them offers they couldn't refuse. As one anonymous dot-com executive remembered AOL's tactics, "For weeks it was, 'You're great, you're great, you're great,' and then one day [we had to] give them every last dollar we had in the bank and 20 percent of our company." Another dot-commer said AOL demanded 30% of her company, "and then for good measure they tell us, 'These are our terms. You have 24 hours to respond, and if you don't, screw you, we'll go to your competitor.' "[30]

In essence, AOL leveraged its "platform" of eyeballs and dial-up customers in the same way that Microsoft had leveraged its operating system. And burnishing this reputation as the 800-pound gorilla of the dot-com market was very lucrative for the company. In the era of skyrocketing valuations, no other Internet company soared as high as AOL did. Over the course of the 1990s, AOL's stock appreciated

80,000%.[31] By 1999, its market cap would reach $149.8 billion, and that same year AOL became the first Internet company added to the S&P 500 index, taking the place of the century-old Woolworth Corporation.[32] AOL was worth more than Disney, Philip Morris, or even IBM; it was worth more than General Motors and Boeing *combined*.[33]

But the gorilla had a problem.

It was no secret to anyone in the tech industry that the days of dial-up modems were numbered. The long-promised dream of broadband—web browsing at speeds thirty times faster than the 56,000 bits per second that was state-of-the-art for AOL's millions of users—was just around the corner. And the biggest issue for AOL was the inconvenient reality that cable companies were in the best position to deliver this new era of connectivity. AOL had achieved ubiquity by piggybacking on the government-regulated copper wires of the staid, century-old telephone network. Unlike on the phone lines, AOL could not expect to get common carriage on cable networks. AOL's bread and butter—being America's ISP of choice—was careening rapidly toward extinction, and everyone inside the company knew it. By doing deals with almost every player in the space, the gorilla had access to everyone's financials, and could see (even before the press caught on) that many dot-coms were close to running out of money.

So, as the dot-com party lurched to its climax, AOL, more than anyone else, knew it was time to find a seat before the music stopped. Fortunately, AOL had one very big ace in the hole: its soaring stock. It could use its gargantuan market cap to buy another company—any company, but preferably one with valuable long-term assets—in order to make up for the inevitable shortfall that would come when dial-up users jumped to broadband. As early as December 1998, internal AOL emails show that Steve Case and his lieutenants began kicking around the idea of purchasing a safe lily pad to land their company on. AOL came very close to acquiring eBay, but Case was wary of doubling down on the Internet space. A merger with AT&T was floated as a way for AOL to claim direct ownership of distribution pipes, but Ma Bell declined. After approaching Disney and getting rebuffed by its CEO, Michael Eisner, AOL turned its focus to arguably the biggest media company in the world: Time Warner. A deal with Time Warner would allow AOL to marry its new media savvy to the toniest of old-media content. Aside from its numerous, tangible and lucrative assets (magazines, TV channels, movie studios and more) Time Warner had one key piece of the puzzle that AOL craved: the second-largest cable network in the country.

Time Warner, of course, was the one big media company that had taken the Internet seriously from the very beginning—and it had hundreds of millions of dollars in losses to show for it. Its CEO, Jerry Levin, was the man who had bankrolled the expensive, doomed Full Service Network and Pathfinder experiments. In spite of these high-profile failures, Levin remained a true believer in technology's ability to transform the distribution of content.

Especially given the way it all turned out, many have painted the AOL/Time Warner merger as a smash-and-grab job: savvy Internet punks swooping in and taking advantage of clueless old-media types. In some sense, it's hard *not* to see the merger as a cynical ploy for AOL to cash in on its market cap before its business model collapsed. But, from another angle, Steve Case probably made the most rational move on behalf of his shareholders. "We all knew we were living on borrowed time and had to buy something of substance by using that huge currency," one AOL executive said later.[34] "We didn't use the term bubble," said another exec. "But we did talk about a coming 'nuclear winter.'"[35]

And from the perspective of Time Warner? As the great tech journalist Kara Swisher said, if Time Warner got conned, "it was clear it was a con that the victim was very much in on."[36] By 1999, when Internet stocks were worth more than gold, and when new phenomena like Napster were driving home the lesson that web technologies could be existentially threatening to old media companies and their distribution models (more on Napster in a later chapter), how could it have felt like anything other than a coup for Time Warner to team up with the ostensible king of the web? By marrying AOL, Time Warner would insulate itself against the Internet's disruption.

At the time it was announced to the world, the merger of $164 billion AOL and $83 billion Time Warner seemed like nothing less than the triumph of the New Economy. "This is a historic moment in which new media has truly come of age," Steve Case told the stunned financial press. "We are going to be the global company of the Internet age."[37] Case vowed that one day AOL Time Warner would have $100 billion in revenue and a $1 trillion market cap. And for the moment, there was no reason to disbelieve him. For all the talk of the deal being a "merger of equals," AOL shareholders would control 56% of the company and Time Warner shareholders, 44%.[38] The reality was, AOL had bought Time Warner. An Internet upstart had taken over a decades-old media giant with five times its revenue.[39]

The entire business world was shocked by the deal. "Let's be clear," Silicon Valley venture capitalist Roger McNamee said. "This is the single most transformational event I've seen in my career."[40] Music industry executive Danny Goldberg said the merger "validates the Internet and vindicates the value of content."[41] The definitive book on the merger, *There Must Be a Pony in Here Somewhere*, was written by Kara Swisher some years after the deal was consummated. In it, she claims that, at the time, the merger seemed like a home run to her and nearly everyone else: "In one major move, the two companies had seemingly addressed their weaknesses and intensified their strengths. I won't deny I really believed that, as did many others—many of whom now pretend they never did."[42]

The AOL/Time Warner merger was announced on January 10, 2000. On April 3, 2000, Judge Jackson's final ruling suggesting the breakup of Microsoft was announced. At the time, these two events felt epochal—clarion signals ushering in a new era in the technology and even media industries.

Instead, from the perspective of hindsight, they look more like historical footnotes, bracketing the weeks when the dot-com bubble finally burst.

■

FOUR DAYS AFTER the AOL/Time Warner merger announcement, on January 14, 2000, the Dow Jones Industrial Average peaked at 11,722.98, a level it would not return to for more than six years. The tech-heavy Nasdaq peaked on March 10, 2000, at 5,048.62, a level it would not reach again until March 2015. From that March 2000 peak, all the way down to the trough it reached on October 9, 2002 (the bear market bottom would be 1,114.11), the Nasdaq would lose nearly 80% of its value.

Was there any one thing that pricked the dot-com bubble? Of course not. There were a myriad of factors that all accumulated to bring about the end of irrational exuberance. For one thing, the Fed had finally begun to raise interest rates: three times in 1999 and then twice more in early 2000, the most sustained round of fiscal tightening over the whole of the late 1990s. And just as suddenly, the language from the Fed had shifted to an open attempt to rein in equity prices. Added to this was the fact that the Internet cheerleaders were changing their tune as well. One by one, Wall Street analysts began advising their clients to lighten up on Internet stocks, saying that the technology sector was "no longer undervalued."[43] But more than anything else, it was the weak constitution of all those "iffy" dot-coms that had hit the market toward the tail end of 1999 that tipped the scales. These were companies without a realistic chance to make money over the long term. Many, perhaps most, had merely been cynical plays to go public and then hope more money could be raised later to keep them afloat.

The crash had myriad victims, but a few can stand for the many. Webvan burned through more than $1 billion before declaring bankruptcy in July 2001.[44] Pets.com had the ignominious distinction of liquidating a mere 268 days after its February 2000 IPO.[45] It closed its first day of trading at $11, the same price at which it had gone public—no first-day pop. The next week of trading saw it down at $7.50.[46] eToys went out of business after ringing up $274 million of debt. Once valued at $10 billion, its liquidators couldn't even line up bidders for the $80 million warehouse system it had built.[47]

By April, just one month after peaking, the Nasdaq had lost 34.2% of its value.[48] Over the next year and a half, the number of companies that saw the value of their stock drop by 80% or more was in the hundreds. By August of 2001, eTrade was down 84% from its all-time high. SportsLine was down 99% (trading at 91 cents). And for most, no recovery ever came, even for the biggest names. Priceline had cratered 94%. Yahoo was down 97%, from an all-time high of $432 per share to $11.86 on August 31, 2001, its market cap down to $6.7 billion from $93 billion. That $1,000 put into Amazon's IPO, which had climbed in value to more than $61,000 at the bubble's height, was worth about $3,400 at the end of September 2001, when Amazon was trading under $6.

There are various ways to measure the amount of wealth that was annihilated when the bubble burst. As early as November 2000, CNNFN.com pegged the losses at $1.7 trillion.[49] But of course, that would only count public companies. The amount of money lost to dot-coms that went bankrupt before IPOing or getting acquired would push the calculation of losses higher still. Beyond the public companies, it's estimated that 7,000 to 10,000 new online enterprises were launched in the late 1990s, and by mid-2003, around 4,800 of those had either been sold or gone under.[50] Many trillions of dollars in wealth vanished almost overnight. Obviously that amount of money leaving the playing field had to have some effect on the economy overall. The U.S. government would date the start of the subsequent dot-com recession as beginning in March 2001. By the time of the economic shock from the terrorist attacks of September 11, 2001, there was no longer any doubt. That tragic month of September, for the first time in twenty-six years, not a single IPO came to market.[51]

The dot-com era was over.

"If you had asked me two years ago, does this dot-com thing make any sense, I would have said, no, the bubble will burst," George Shaheen, Webvan's CEO, told the *New York Times* shortly before Webvan went under. "But I didn't have any idea of the blood bath that would ensue."[52] Shaheen, whose Webvan stock options were once worth $280 million, saw the value of his paper wealth shrink to $150,000 by the time he quit the company.[53]

Perhaps most emblematic of this epic turn in fortune is the story of TheGlobe.com. Founded by two undergraduates at Cornell University in 1995, TheGlobe was a community site, allowing things such as personal homepages, much like GeoCities, Angelfire and

Tripod did. It had decent early user growth, reaching 14 million hits a month and 30,000 subscribers by 1996.[54] And, most important, it had young, baby-faced, photogenic twenty-something cofounders, Stephan Paternot and Todd Krizelman.

By 1997, the site was adding 100,000 users per month.[55] This sort of growth attracted the attention of Alamo Rent A Car founder Michael Egan, who invested $20 million. Paternot and Krizelman moved the company's operations to New York City and plunged into the hype-and-party machine that was already in full swing. By late 1998, with revenues of only $2.7 million (and losses of $11 million), TheGlobe.com seemed to be another promising dot-com, ready for its time in the spotlight.

TheGlobe enjoyed what was perhaps the quintessential IPO of the dot-com era. Going public on Friday, November 13, 1998, priced at a cautious $9 a share, Bear Stearns, the underwriters of the stock, discovered that there was suddenly a 45-million-share demand for the 3 million shares TheGlobe was selling.[56] When the stock opened in the morning, the first trade took place at $87 a share. TGLO reached an intraday high of $97 before closing at $63.50. It was the largest single-day IPO pop in history—605.6%. Sixteen million shares traded hands during the day, meaning the 3 million shares of TGLO available to the public were bought and sold on average five times that day. This was, of course, the smart money selling out. "I sold my TGLO at 88—who wouldn't?" one hedge fund manager told TheStreet.com.[57] The headline in the *New York Post* would later read, "Geeks Make $97 Million."[58] Paternot and Krizelman were twenty-four years old.

TheGlobe had executed on the dot-com era playbook perfectly. Except . . .

On the second day of trading, TheGlobe.com's stock fell to $48. Within a week, it was down to $32.

Over the course of 1999, TGLO would rise and fall with the rest of the Internet stocks, briefly bouncing to almost $80. But, quite literally, it was all downhill from there. Toward the end of 1999, the price was down, under $10.

TheGlobe was—perhaps from the beginning—a company with

dubious long-term prospects. For a company that needed gobs of traffic to ever have a chance of making money, it never really competed with the big boys, only peaking at 34 in the list of the most trafficked websites in the world.[59] Because of this "second-tier" status, TheGlobe never had the chance to be acquired like GeoCities and even Tripod were. In an era when an electronic greeting card company like BlueMountain could be snapped up for its 9 million unique visitors a month, TheGlobe was only averaging 2.1 million.[60] Plastering banner ads at the top of every page in order to make money was never a sustainable strategy, especially when the online ad market began to crash by the end of 1999. GeoCities and Tripod were safe under their parent company umbrellas, so who would ever know if they were just as unprofitable? Meanwhile, at publicly traded TGLO, the whole world could see that, even in a good quarter, like the three months ending March 31, 2000, when TheGlobe.com saw revenues more than double, to nearly $7 million, it nonetheless recorded a net loss of $16.4 million.[61] Just by being in business, doing the thing its business plan said it had intended to do, TheGlobe was losing more than $2 for every $1 it brought in.

There are plenty of people, both today and at the time, who view TheGlobe as designed merely to IPO, make its investors and bankers rich, and then—nothing more. Whether or not that was the case, for one brief, shining moment, it was the hottest stock, the most exciting company in the world.

And then, it was a laughingstock.

By the spring of 2001, TheGlobe.com was trading at 8 cents a share. Paternot and Krizelman were forced out of their own company long before then.

When TheGlobe.com was delisted from the Nasdaq on April 23, 2001, its final trading price was 16 cents.

■

OF COURSE, THE DOT-COM ERA didn't end disastrously for everyone. According to numbers subsequently published by *Barron's*, between September 1999 and July 2000, insiders at dot-com companies cashed out to the tune of $43 billion, twice the rate they had sold at during 1997 and 1998.[62] In the month before the Nasdaq peaked, insiders were selling twenty-three times as many shares as they bought.[63] The most famous example from this era is Mark Cuban, perhaps the quintessential dot-com billionaire. Cuban had already cashed out early by selling his company, Broadcast.com, to Yahoo. But he didn't trust the insane valuation of the Yahoo stock he had been paid in, so he set up a hedge against his Yahoo holdings, called an "options collar." When Yahoo's stock subsequently collapsed, his entire fortune was protected. "He probably extracted more from the initial Internet bubble than anyone else," the hedge fund manager and author James Altucher said of Cuban.[64]

Compare Cuban's story to that of Toby Lenk, founder of eToys, who saw his paper fortune of $600 million wither away because he refused to bail out on his company's stock.[65] Is there any great nobility in Lenk's determination to go down with his ship versus Cuban's astute decision to get out when the getting was good? Probably not. Or consider Paternot and Krizelman, who in May of 1999, when TheGlobe stock was still at $20 a share, sold 80,000 shares and 120,000 shares for roughly a combined $4 million (original investor Michael Egan sold TheGlobe shares worth more than $50 million).[66] Paternot, Krizelman and Egan did nothing ethically or legally wrong. In fact, they played by the rulebook of a crazy game that was largely foisted upon them. But one does wonder about other people who had a stake in TheGlobe.com. Those hundreds of employees, say, who had been granted stock options and imagined themselves to be rich the day of TheGlobe's IPO. Or what about the potentially tens of thousands of investors who bought shares of TheGlobe.com for $87 on IPO day? When did they sell? And at what price?

What we can say definitively is that we know who ended up holding the bag as the bubble exploded: average investors. Over the course of the year 2000, as the stock market began its meltdown, individual investors continued to pour $260 billion into U.S. equity funds. This was up from the $150 billion invested in the market in 1998 and $176 billion invested in 1999.[67] Everyday Americans were the most aggressive investors in the dot-com bubble[68] at the very moment the bubble was at its height—and right at the moment the smart money was getting out. According to *Barron's* journalist Maggie Mahar, by 2002, 100 million individual investors had lost $5 trillion in the stock market. Bloomberg News has estimated the damage at $7.41 trillion.[69] A Vanguard study showed that by the end of 2002, 70% of 401(k)s had lost at least one-fifth of their value; 45% had lost more than one-fifth.[70]

A lot has been made in the last several years about income inequality and how gains made in the overall economy tend increasingly to go to the top 1%, while the rest are left with scraps. At the time of this writing, there is a lot of talk about how the American public, especially the middle and working classes, have come to believe the economic structure of America is rigged against them, and everything is tilted in favor of the insider, the moneyed, the elite. An argument can be made that this was a belief that first took hold when the dot-com bubble burst, especially to a generation of investors who came to the stock market for the first time in those years. Baby boomers did what society told them: they invested in stocks; they bought and held. And for a time, they did well, seeing their nest eggs go up by five, even six, figures (or more if they were lucky). And then they watched it all evaporate. They watched the insiders and the bankers, the lucky and the elite, walk away scot-free while they, the hardworking Americans who did what they were told, lost everything. And all of that would happen to them *again* less than a decade later, only this time, in the housing market.

The bursting of the dot-com bubble was the opening act of our current economic era, and the repercussions from that bubble's aftermath are still with us today, economically, socially, and especially politically.

Middle-aged investors weren't the only ones to lose out, of course. A whole generation of workers who had staked their careers on the transformative dream of technology were suddenly, almost en masse, unemployed. It was later estimated that between 2001 and early 2004, Silicon Valley alone lost 200,000 jobs.[71] A whole generation of young people had, in the space of a decade, gone from being young upstarts who "got it," to masters of the universe who seemed to be transforming the world, to completely redundant.

There were some engineers and secretaries at various companies who were lucky enough to cash out some stock options at the right time and probably walked away with enough money to pay off student loans, put a down payment on a house, or maybe pocket a cool million or two. But those were the early or the lucky. The vast majority, the tens or maybe hundreds of thousands who flooded into tech in the bubble era, now found themselves without even a severance package because their pre-IPO company was bankrupt.

The hangover from this comeuppance is what still haunts the tech industry today. Even now when young entrepreneurs talk glowingly about how their technology will change the world, in the back of any Internet entrepreneur's mind is the Icarus-like cautionary tale of the dot-com bubble's implosion, as well as a fear that someday they too will be exposed for their hubris.

Marc Andreessen would later say of the bubble and its aftermath: "A lot of big companies in 2000, 2001, 2002, breathed a massive sigh of relief, and said 'Oh! Thank God that Internet thing didn't work! Stick a fork in it, it's done. Everybody knows the dot-com thing was a bubble. That was a joke. It's over. So, now we don't have to worry about it.' "[72]

American industry need look no further than the example of AOL Time Warner to assure themselves that it had all just been a grand delusion. A multitude of books have been written about the monumental culture clash that ensued when the AOL cowboys invaded the halls and boardrooms of Time Warner. Certainly, dysfunctional infighting and managerial malfeasance were in large part responsible for what conventional wisdom has collectively agreed was the worst merger of all time. There were charges of accounting fraud and dirty dealing directed at the AOL side, but the ultimate failure of the combination was really a result of the collapse of AOL's advertising business. Over the course of 2000 to 2002, all those deals AOL did with dot-com companies unwound, as the dot-coms themselves went belly-up. Slowly, AOL dial-up subscriber numbers, which peaked at 26.7 million in 2002, dwindled away, as Americans shifted over to broadband connections with DSL companies or to cable ISPs like Time Warner Cable's own Road Runner Internet service.[73] As early as 2003, Time Warner dropped "AOL" from its corporate name.[74] By that point, Steve Case and most of the rest of the AOL cowboys had been pushed out of the company. But also by that point, AOL Time Warner had been forced to announce two of the biggest losses in American history: $54 billion in 2002 and $45.5 billion in 2003, both write-downs of the inflated value of AOL's market cap that was now proven to have been illusory.[75] Corporate America assured itself that it had been right all along: there was little money to be made on the Internet, and the evaporation of the biggest Internet player of them all seemed proof positive.

■

MANY OBSERVERS of the dot-com bubble have found it instructive to compare it to earlier bubbles like the tulip mania in seventeenth-century Holland or the South Sea Company's collapse in eighteenth-century London. But it's the example of the railroads in Britain in the 1840s that's the most analagous.

Railways were cutting-edge in the 1840s. As with the dot-coms, there was a period of about three or four years when Britons experienced a mad rush to invest in business schemes surrounding this new technology. Eight hundred miles of new railways were floated for development in 1844. Two thousand eight hundred and twenty miles of new track were proposed in 1845. A further 3,350 miles were authorized in 1846. Because the British Parliament had to pass legislation approving each new railway scheme, the railway bills passed by Parliament provide an amusing analogy to the IPOs of the dot-com period. Forty-eight railway acts were passed by Parliament in 1844, and 120 in 1845. At the height of the mania, the capital required to fund these schemes came to £100 million, and by 1847, investment in the railways represented 6.7% of all national income.[76]

In his book *Fire and Steam: A New History of the Railways in Britain*, historian Christian Wolmar describes a frenzy that sounds eerily familiar:

> As the supply of finance appeared almost endless, with more and more people eager to jump on the "get rich quick" bandwagon, unscrupulous fraudsters entered the fray, pushing schemes whose only aim was to deprive investors of their savings. For example, investors were being sought for schemes whose sole purpose was to pay the bills on previous projects drawn up by the same promoters. While such utterly fraudulent schemes were few, there were many more in which investors lost their money because the economics were as shaky as their prospectuses were woolly.

The inevitable bust came because, in Wolmar's words, the bubble was ultimately based on "little more than optimism feeding on itself," and it was pricked in part by the Bank of England raising interest rates.[77] The aftermath of the bubble feels similar to the aftermath of the dot-com fiasco, albeit with a Victorian tinge:

> A contemporary chronicler reckoned "no other panic was ever so fatal to the middle class. . . . There was scarcely an important town in England what [*sic*] beheld some wretched suicide. It reached every hearth, it saddened every heart in the metropolis. . . . Daughters delicately nurtured went out to seek their bread. Sons were recalled from academies. Households were separated; homes were desecrated by the emissaries of the law."[78]

But what Wolmar's account also points out is to what degree the bubble, and the railroads constructed because of it, ultimately created the infrastructure that would enable the high Industrial Revolution in Victorian Britain. The mileage of rail schemes authorized during the bubble years came to represent 90% of the total route mileage on Britain's rail system. "The vast majority of the railways constructed in these years survive today as the backbone of the [UK rail] network," Wolmar writes.[79]

The bubble made possible the British Empire at its economic height. People never stopped riding trains. Businesses never stopped shipping goods over them. The railways never went away, even after the investment mania did. The lesson of the dot-com bubble is similar. Of course, the dot-coms went away. Of course, AOL—for one brief shining moment, the embodiment of the Internet in American life—went away. But the *Internet itself* didn't go away. And that's why the railway example is so pertinent.

All of the money poured into technology companies in the first half decade of the Internet Era created an infrastructure and economic foundation that would allow the Internet to mature. And I mean that in a tangible, physical way. During the dot-com bubble, there was a similar, less publicized bubble in telecommunications companies. *This* estimated $2 trillion bubble ended in a similar bloodbath with the well-publicized bankruptcies of companies like WorldCom and Global Crossing.[80] But before the bubble burst, between the years 1996 and 2001, telecom companies raised $1.6 trillion on Wall Street and floated $600 billion in bonds to crisscross the country in digital infrastructure (the banks collected more than $20 billion in fees for their troubles, far more than they had gotten from the dot-com IPOs).[81] These 80.2 million miles of fiber optic cable represented fully 76% of the total base digital wiring installed in the United States up to that point in history.[82] What did this mean, ultimately? Well, it meant that for the coming years, the literal infrastructure that would allow for the maturation of the Internet was in place. And because of a resulting glut of fiber (the telecoms had overextended themselves just as disastrously as the dot-coms, thus the bankruptcies) in the years after the dot-com bubble burst, there was a severe overcapacity in bandwidth for Internet usage that allowed the next wave of companies to deliver sophisticated new Internet services on the cheap. By 2004, the cost of bandwidth had fallen by more than 90%, despite Internet usage continuing to double every few years.[83] As late as 2005, as much as 85% of broadband capacity in the United States was still going unused.[84] That meant as soon as new "killer apps" were developed, apps like social media and streaming video, there was plenty of cheap capacity allowing them to roll out to the masses. The tracks, as it were, had already been laid.

And people didn't suddenly stop surfing the web. Many have made the case that the dot-com era was doomed to failure simply because there were too many companies chasing what at the time were too few users. When the bubble burst in 2000, there were only around 400 million people online worldwide. Ten years later, there would be more than 2 billion (best estimates peg the current number of Internet users at 3.4 billion).[85] In the year 2000, there were approximately 17 million websites. By 2010, there were an estimated 200 million (today, that number is over a billion).[86] In 2000, a company like Yahoo could claim a $128 billion market cap because it was tallying 120 million unique visitors a month.[87] A decade later, Yahoo would boast a global monthly audience of 600 million.[88] Amazon might have flirted with insolvency after the bubble burst, but the company has seen its revenue increase every year of its existence, even in the worst years of the bubble's aftermath. Amazon's revenue in 2000 was $2.8 billion. Ten years later, it would be $34.2 billion.[89]

Far from being a fad, the habits Americans acquired during the bubble era ingrained themselves into the rhythms of everyday life. The dot-coms, the training wheels for the Internet, the pioneers, they all taught us to live online. We all might have jumped from dial-up to broadband, but few of us quit using the net. There was no going back.

And even as the dot-com companies were crashing and burning, there were already new innovators on the scene who would move the Internet forward in an entirely new, entirely personal, and (finally) exceedingly profitable way.

11

I'M FEELING LUCKY

Google, Napster and the Rebirth

When Larry and Sergey first met, they didn't like each other much.

In the summer of 1995, Larry Page was considering a transfer to Stanford University's graduate program in computer science. Sergey Brin was already two years into the program, and he had signed up to be a tour guide of sorts to potential students. One summer day, he showed Page and a group of other potential Stanford students around the Bay Area.

"I thought he was pretty obnoxious," Page said later of his guide. "He had really strong opinions about things and I guess I did, too."

"We both found each other obnoxious," Brin agrees. They might have stepped on each other's toes a bit, but at the same time there was a degree of frisson to the encounter. "We spent a lot of time talking to each other," Brin would recall, "so there was something there. We had a kind of bantering thing going."[1]

On the surface, it might not have seemed like Page and Brin had anything in common. Page was a midwesterner, born in East Lansing, Michigan, on March 26, 1973. Brin was born in Moscow, in the Iron Curtain–era USSR, on August 21, 1973, and was brought to the United States when he was six years old. Page was reserved, quiet, contemplative. Brin was outgoing, gregarious, loud. Page was a deep thinker, a visionary. Brin, a problem solver, an engineer's engineer.

But the two had more in common than anyone knew that first day. They both came from academic families. Page's father was a pioneering computer science professor at Michigan State University, where his mother was also a computer programming instructor. Brin's father was a mathematics professor at the University of Maryland and his mother a researcher at NASA's Goddard Space Flight Center. Larry and Sergey both grew up to respect research, academic study, mathematics and especially computers. And they both had inquisitive minds, believing in the power of knowledge to overcome any obstacle, intellectual or practical. Each had been inculcated into this spirit of intellectual fearlessness at a young age.

"You can't understand Google," early Google employee Marissa Mayer (and later, Yahoo CEO) has insisted, "unless you know that both Larry and Sergey were Montessori kids. It's really ingrained in their personalities. To ask their own questions, do their own things. Do something because it makes sense, not because some authority figure told you. In a Montessori school, you go paint because you have something to express or you just want to do it that afternoon, not because the teacher said so. This is baked into how Larry and Sergey approach problems. They're always asking, why should it be like that? It's the way their brains were programmed early on."[2]

For Larry and Sergey, their intellectual fearlessness overlapped in such a way that their conflicting personalities actually ended up complementing each other. When Page came to Stanford for the 1995–96 academic year, he and Brin became close. Friends took to calling the duo LarryandSergey, and the pair would end up debating endlessly on topics ranging from philosophy to computing to films, two equally matched polymaths thrilling to the intellectual joust. Brin's hobby project was creating a software program that could provide movie recommendations based on the tastes and viewing habits of other people who had seen similar films (not unlike what Netflix later perfected). Page's dream obsession was creating a system of networked, autonomous cars to ferry people around.

Even though they were the same age, Brin was academically two years ahead of Page because he had completed his undergraduate computer science degree at age nineteen and aced all of Stanford's required doctoral program exams on the first try.[3] But despite this head start, and despite being the recipient of a National Science Foundation fellowship that allowed him to do basically anything he wanted, Brin had stalled out in his quest to nail down a dissertation topic. Of course, the newly arrived Page also needed to decide on *his* dissertation, and so fate pushed the pair even closer together. In January 1996, LarryandSergey ended up working in the same office, number 360, in the just-completed William Gates Computer Science Building on Stanford's campus. The building was of course named after the founder of Microsoft, who had donated $6 million to the construction. All his career, Bill Gates repeatedly predicted that one day, some student somewhere would found a company that would challenge Microsoft for dominance of the tech industry. His prediction turned out to be right, and from a building with his name on it.

▪

PAGE WAS STRUCK by a fundamental truth about the web that is glaringly obvious when you state it out loud: it is built on links. One page linking to another; one idea linking to another. As of yet, no one had bothered to analyze the structure of the link ecosystem in a comprehensive way. For example, it was possible to know that webpage A linked to webpage B because you could see it—you could follow the link. But what about the reverse? What pages had linked webpage A? There was no way to know. You couldn't follow a link stream backward, only forward. Page wondered: if you analyzed all of the back links, if you mapped out the link structure of the entire web, what sort of insight might that data give you?

Page's intuition was that this might be more than just an interesting theoretical question. As he mulled over the idea with Brin, their shared upbringing as the children of academics kicked in. LarryandSergey knew the power of the academic citation. Their parents had published academic papers. They, themselves, intended to publish academic papers in order to earn their Ph.D.'s. And they knew that any academic paper worth its salt built its argument by citing other academic papers and studies. In the world of academia, those citations, the accumulated number of "votes" from paper to paper, served, over the years, to accrue value to given ideas—to essentially rank them based on the number of citations. The most cited papers were understood to be the most authoritative. "It turns out, people who win the Nobel Prize have citations from 10,000 different papers," Page would say later.[4]

Well, what was a web link but a digital citation? If you analyzed the links, analyzed the citations, you might be able to make inferences about the relative value of a given web page, and possibly even determine which webpage was more authoritative by analyzing the back links in the same way that counting the citations told you which academic paper was the definitive one. Larry Page wanted to map out the value of the web's connections by going backward through the link chain. Page went to his academic advisor, Terry Winograd, and asked for the money and machines that would allow him to map the web's links. He dubbed the project BackRub. When asked how much of the web he intended to map, he replied: "the whole web."[5]

So, in March of 1996, Larry Page launched BackRub by sending search bots, known as "spiders," out into the web to find *all the links*. He started with a single page—the Stanford computer science department homepage—and then fanned out, following link after link, cataloging them all, and then ranking web pages based on these link citations. It was the mathematical complexity of this ranking—the complicated problem of determining which page was more valuable based on a combination of accumulated links as well as the authority passed through from pages that linked to other pages—that drew Sergey Brin to join the project. Larry and Sergey called their combined citation-ranking system PageRank, either as an ode to Page himself or as an obvious descriptor of what the system was intended to do.

"The idea behind PageRank was that you can estimate the importance of a web page by the web pages that link to it," Brin says. "We actually developed a lot of math to solve that problem. Important pages tended to link to important pages. We convert the entire web into a big equation with several hundred million variables which are the PageRanks of all the web pages, and billions of terms, which are all the links."

"It's all recursive," Page said. "In a way, how good you are is determined by who links to you and who you link to determines how good you are. It's all a big circle."[6]

LarryandSergey suddenly had a project that would make for a pretty interesting dissertation. And as soon as the pair looked at their results, they realized their intuition was dead on: the citation analogy worked. If you wanted to find the most authoritative webpage about a topic such as, say, windsurfing, BackRub/PageRank could tell you. It would know based on the accumulated links, of course, but also from the authority passed on from other authoritative sites. Thanks to Brin's math (largely linear algebra and something about the eigenvector of a weighted link matrix, for those who know what that means), citations from obviously important websites were more valuable than others. A link from some unknown person's personal webpage might be valuable, but a link from a professional windsurfer would be judged to be even *more* valuable— and a link from, say, Yahoo's homepage would be even more valuable still.

It was at this point that the really interesting application for this little math project became obvious. "It was pretty clear to me and the rest of the group," Page said later, "that if you have a way of ranking things based not just on the page itself but based on what the world thought of that page, that would be a really valuable thing for search."

■

IT TURNED OUT THAT the reason search engines had never worked very well prior to PageRank was not that they were broken, but because they were missing the key innovation that Brin and Page had stumbled upon: relevancy. If, in 1997, you did a search for "automobile company" on even the best search engine at the time (AltaVista) you'd find yourself disappointed because the websites of

Ford, General Motors or Toyota would probably not show up. It's not that AltaVista couldn't find those sites. It most certainly had! Ford.com or GM.com or Toyota.com were most likely in the list of tens of thousands of results that AltaVista had found. It was just that AltaVista had no way of surfacing those most relevant results to the top. So they were on page 3 of the search results. Or page 300.

PageRank solved this problem. PageRank knew which sites were the most authoritative automotive sites already, and so when you combined its algorithmic prowess with the traditional tricks of information retrieval that all the search engines were already using, suddenly it all just worked. Indeed, as Page and Brin combined BackRub and PageRank with traditional search methods like analyzing on-page text, webpage titles or metatags and, especially, parsing the so-called anchor text of a link (someone who makes a link out of the words "flower shop" and then points it to a given website is really trying to tell you something), they realized PageRank was incredibly powerful. Page and Brin discovered that their algorithm was indeed recursive, meaning that the more data they fed it, the more webpages it analyzed, the better it got. By tweaking the math even more, LarryandSergey's search tool could reliably find people, locate the most obscure fact or data, and even *answer questions*. PageRank wasn't finding new things. It was merely finding things in a better way. The earlier search engines were already answering every query correctly. But it was finding the needle in the haystack and putting it at the top of the list that PageRank did better.

"It wasn't that they [Page and Brin] sat down and said, 'Let's build the next great search engine,'" said Rajeev Motwani, who was Brin's academic advisor. "They were trying to solve interesting problems and stumbled upon some neat ideas."[7]

■

IT WAS A GOOD THING Page and Brin had not set out to build the next great search engine, because, at the time, no one was really clamoring for one. In the late nineties, when Page and Brin began refashioning BackRub/PageRank into a search engine, there was a universe of major search players: Yahoo, Excite, Lycos, AltaVista, AskJeeves, MSN, and on and on. In a time when Yahoo had a $100 billion market cap, who needed another entrant into an already-crowded space, no matter how superior it was? Fortunately, Page and Brin were not business-focused at that time. They were academics, more interested in defending a dissertation and publishing a paper on their research than starting a company around their idea.

So, they produced that paper: "The Anatomy of a Large-Scale Hypertextual Web Search Engine," which was presented at a conference in Australia in May of 1998. But if Page and Brin initially stayed true to their chosen academic paths, that did not mean they were blind to the financial possibilities inherent in their work. How could they have been? Students studying computer science in the heart of Silicon Valley couldn't help but notice what was going on all around them. "It was a hard time to stay in grad school," remembered Tamara Munzner, one of the students sharing room 360 of the Gates Building with Page and Brin. "Every time you went to a party, you had multiple job offers and they were all real. I had to redecide every term not to leave."[8]

The obvious move was to license PageRank to one of the existing players, and indeed, this is what Page and Brin attempted to do. They met with everyone from the Yahoo founders Jerry Yang and David Filo, to another search pioneer, Infoseek's Steve Kirsch. No one was interested. The closest they came to making a deal was when Page wrote up an extensive proposal to Excite's leadership, suggesting they replace Excite's existing algorithms with his. Doing so, he calculated, would generate an additional $47 million in revenue for the search engine. "With my help," Page wrote in his proposal, "this technology will give Excite a substantial advantage and will propel it to a market leadership position."[9] All he asked for in exchange was a seemingly reasonable $1.6 million in cash and Excite stock—a nice little payday—and then he and Brin would return to finishing their doctorate work. Excite countered with $750,000, which Page and Brin rejected.

The incumbent search players' failure to scoop up the PageRank technology has become infamous in business lore as one of the great missed opportunities of all time. Larry Page has, on a few occasions, suggested that the search companies were simply myopic. "They were becoming portals. We probably would have licensed it if someone gave us the money. . . . [But] they were not interested in search," Page has said. "They did have horoscopes, though."[10] But Excite CEO George Bell has a slightly different recollection: "The thing that Larry insisted on, that we all do recall, is that Larry said, 'If we come to work for Excite, you need to rip out all the Excite technology and replace it with [our] search.' And, ultimately, that's—in my recollection—where the deal fell apart."[11] This was Page and Brin's intellectual fearlessness demonstrating itself for the first time in a competitive setting. The pair believed—*knew*—that they had a superior way of doing things, and so they thought nothing of going to an established search company and telling them their existing product sucked. This brashness had the effect of insulting Excite. Excite was a company founded by brilliant Stanford computer scientists, after all. "We had hundreds of engineers at that point," Bell points out. Why should the company furlough their engineers just because two other engineers had come along with claims to be *more* brilliant? Bell claims that there was no way he could justify upsetting his existing talent, especially when some of them were founders of the company. "Ultimately I couldn't stomach the cultural risk that Larry insisted on," Bell says.

But if Page and Brin were confident almost to the point of being arrogant, they certainly had plenty of data to back them up. In order to fine-tune their algorithm, the pair had needed plenty of real-world feedback. Starting in 1997, they had made the search engine available, first on Stanford's internal network, and then to the general public. Through nothing but word of mouth, the service grew increasingly popular, serving more than 10,000 queries a day by late 1998.[12] Page and Brin monitored the server logs and made tweaks to their system based on the data this provided. They named the service Google, a play on the word "googol," which is a 1 followed by 100 zeros. The idea was to suggest they were capturing the whole web, everything in existence. "The name reflected the scale of what we were doing," Brin said later.[13] Googol.com was not available, so Google.com became the URL of the public service.

The popularity of the service, combined with the vast computing resources eaten up by the spidering and indexing, meant that the Google project was rapidly outgrowing the scope of a simple research project. Even when it was housed on a single machine in a Stanford dorm room, Google was hogging large amounts of the university's bandwidth. Stanford was, as ever, incredibly accommodating to an idea born within its walls, but the institution's generosity had a practical and obvious ceiling.

It was clear that if they wanted the Google experiment to continue, Page and Brin would need more resources. More computers, more bandwidth, more people to work on the algorithm—this all meant more money than a research budget, even a generous one, could provide. So the pair turned to another Stanford faculty advisor, David Cheriton. Cheriton introduced the pair to Andy Bechtolsheim, a successful entrepreneur who had founded Sun Microsystems while also a Ph.D. student at Stanford. One morning in late 1998, Page and Brin met Bechtolsheim at Cheriton's home. Bechtolsheim made out a check on the spot for $100,000 in the name of Google Inc. The check sat in Page's dorm room desk for a number of weeks before Google Inc. was formally incorporated on September 7, 1998. Page and Brin would raise an additional $1 million when David Cheriton kicked in some money, as well as a few others, including former Netscape executive Ram Shriram and Jeff Bezos of Amazon.

Page and Brin were now entrepreneurs, if perhaps still a little reluctantly. But they were not entrepreneurs in the mold of so many others in the dot-com era. Rather than blowing Google's funds on lavish launch parties or marketing campaigns, they stayed grad students at heart, and instead invested all the money they had raised in continuing their project efficiently. Instead of building out their system by buying software from Microsoft, they used the free Linux operating system. Instead of splurging $800,000 on setups from IBM or Oracle, they spent a mere $250,000 to cobble together a rack of eighty-eight computers to meet their number-crunching needs. At

Stanford they had begged, borrowed, and almost quite literally stolen the computers they needed to keep Google running. Now, they simply switched to buying computers off the shelf from Fry's, the famous Silicon Valley electronics store, and fashioned them into a strung-together system of their own design. Part of this was simple frugality, a habit that would serve them well when the dot-com bubble burst a few short years later. But a lot of it was Page and Brin's ingrained Montessori philosophy: they never met a problem they couldn't solve through smart engineering.

Google didn't take pages from the established Silicon Valley playbook because, in a way, they had never bought into it. They didn't try to Get Big Fast. Instead, Page and Brin were almost manically focused on endlessly iterating and improving upon their Big Idea, making sure it was the most comprehensive, reliable and—most important—speedy search engine in the world. Nothing Google did in its first years distracted the company from improving on its core product. This confidence that they could do everything better proved, in the coming years, to be something of Google's secret sauce. Not only did Google's search engine continue to be superior to any rival in existence, it slowly but surely widened the gap between its version of search and the competition. And their frugality paid off in efficiency. Some observers estimated that "for every dollar spent, Google had three times more computing power than its competitors."[14]

Frugality and efficiency were not just virtues, they were also philosophical and aesthetic differentiators. Google's home page was simply the Google logo, a text field to enter a search query, a search button to execute that query and a button that said I'M FEELING LUCKY, which automatically took you to the first result returned. If you went to the search results page, you only got a list of links. And that was it. No ads, no banners, no weather, no stock quotes, no horoscopes. All the rest of the page was just copious white space. In an age of portals where every other search site was a sea of distractions meant to keep you from, you know, getting to the page you were looking for, Google stood out from the crowd with its single-minded purpose and simplicity. By keeping the pages to almost exclusively text, Page and Brin could ensure they loaded quicker than the search pages of their competitors, and expensive processing power wasn't wasted loading graphics.

This all paid dividends many times over in Google's steady growth. By 1999, usage of the search engine was increasing by as much as 50% a month.[15] From 100,000 searches a day at the beginning of that year, Google searches grew to an average of 7 million per day by the end of it.[16] Overall traffic to the Google homepage was peanuts compared to the numbers a site like Yahoo was pulling down, but in the case of Google, its users came via word of mouth alone. Not a dime was spent on marketing or promotion. Rave reviews from the media continued to turn people on to the service. The *New Yorker* said Google was "the default search engine of the digital in-crowd."[17] *Time* Digital said: "Google is to its competitors as a laser is to a blunt stick."[18] Ordinary users simply told one another about how great and useful Google was. More often than not, users would become Google converts for life.

An early article on Google in *Fortune* from November 1999 summed up a new user's experience. Describing the site as "inscrutable magic," journalist David Kirkpatrick offered this anecdote. On the occasion of the 1999 American League playoffs, Kirkpatrick typed "New York Yankees 1999 playoffs" into both Google and Alta Vista. "The first listing at Google took me directly to data about that night's game," Kirpatrick wrote. "The first two at Alta Vista linked to info about the *1998* World Series." Only by clicking the third Alta Vista link, and then visiting an additional link, did he find the information he was originally searching for. Kirkpatrick's conclusion: "Google really works."[19]

In that same article, Sergey Brin was quoted as boasting, "We're building a way to search human knowledge." If Google was meant to organize all the information in the world, it would need resources on an industrial scale. That same brashness continued to manifest itself when Google needed to raise yet more money.

Despite the glut of search companies already on the market, Google had gotten the attention of venture capitalists, and they were ready to invest in these refugees from academia. But, confident as ever, Page and Brin gave off the impression that they didn't need anyone's help or money. In meetings with potential backers, the pair refused to divulge even basic details about how their service was operating. Their stonewalling even led one prominent VC to storm out of their office in anger. "Larry and Sergey didn't have the language to say things nicely," recalled Salar Kamangar, an early employee who bore witness to Google's general evasiveness during the fundraising process. "They'd be kind of blunt and say, 'We can't tell you.' And the VCs would get very frustrated."[20] The truth was, Page and Brin did not want to take money from just any old VC. They only wanted the best: Kleiner Perkins and Sequoia Capital. The pair proposed that each firm, the blue chips of Silicon Valley venture, take a coequal stake in Google. There was usually one "lead" investor in a round of startup financing, and both KP and Sequoia had enough clout on their own that they had never before deigned to share the spotlight with another firm.

Page and Brin wanted the firms to split the round because that would allow them, as the founders, to maintain a majority share in the company, and thereby retain control of their own destiny. They even had the temerity to issue an ultimatum: each firm would invest $12.5 million in Google, for a total of $25 million, take it or leave it. On June 7, 1999, the VCs took the deal, and Kleiner's John Doerr and Sequoia's Mike Moritz joined Google's board of directors. The only concession the money men had been able to wring out of Page and Brin was a promise to hire someone experienced to take over as CEO of the company at some point in the near future.

This huge round of financing not only put Google firmly on the technology world's map, it went a long way toward ensuring the company's long-term survival. This war chest of money, coming just before the dot-com bubble burst, combined with Larry and Sergey's frugal ways, meant that Google would survive the coming nuclear winter. Had Google waited a further year to raise money, it might not have been able to. And by virtue of being flush with cash when the rest of Silicon Valley was seemingly going belly-up, Google was able to have its pick of talent when the dot-com layoffs began.

Just as it had been frugal when others were profligate, Google also bucked prevailing dot-com habits when it came to hiring. The company put off drafting an army of sales and marketing people until much later. Instead, in 1999 and 2000, Google staffed up with—what else?—brainiacs. Larry and Sergey hired software engineers, hardware engineers, network engineers, mathematicians, even neurosurgeons. Just as with every other facet of their company, Page and Brin wanted only the very best. They wanted Ph.D.'s and scientists. Google would become notorious for the rigorous way it interviewed and screened potential hires—and for its exacting selectiveness. For many years, every new employee was personally vetted by Brin and Page themselves, who expected candidates to measure up to their own intellectual standard. "We just hired people like us," Page said.[21]

Google was able to attract talent because it was nothing short of beloved in Silicon Valley. Here was an Internet company that had solved a universally recognized problem through smart thinking alone. This created a reputational halo that was only enhanced by Larry and Sergey's increasingly bold and public enunciation of Google's mission, which was eventually formalized as an attempt "to organize the world's information and make it universally accessible and useful." While so many dot-com companies claimed to be changing the world by offering dog food online, here was a company that truly seemed revolutionary in the most expansive sense of that word. "Ultimately I view Google as a way to augment your brain with the knowledge of the world," Sergey Brin said.[22] It helped that Google positioned itself as the anti–dot-com startup. Glitz, hype and excess were out; frugality, hard work and earnestness were in. And when Google came up with its famous motto (Don't Be Evil) everyone in technology read between the lines and believed that Google was staking a claim to be the anti-Microsoft.

Google did pick up a few habits from its dot-com brethren, but in typical Larry and Sergey fashion, it did so with a twist. By the time Google moved to its first truly professional digs—an office park in Mountain View that would be dubbed the "Googleplex"—a system of perks for Google's workers were put in place, but they were instituted with an eye toward productivity. The food in the cafeteria was always free, with an in-house gourmet chef; private bus lines picked up workers from around the Valley to shuttle them to work; masseuses roamed the hallways; there were free fitness classes and gyms; and on and on. But every one of these perks was self-consciously provided as a way to keep workers motivated and productive. The free cafeteria meant that Google employees didn't have to leave the office in the middle of the day and could get back to work with ease. In the bathroom stalls were quizzes and coding tips to help people stay sharp. The shuttle buses had WiFi on them, so employees could be productive on the way to and from the Googleplex. Healthy, clear-headed workers could do better coding, or so the thinking went.

All of this combined to make Google *the* technology company to join right as the dot-com bubble burst. If you got hired at Google, it elicited envy from your peers not only because they felt you were doing the most interesting work in technology, but because it meant you were among the best and the brightest. Anyone could get hired at a dot-com toward the end of the decade. But not everyone—even the smartest of the smart—could make the cut at Google. And when the bubble burst and it was seemingly the only company still hiring, the dream of the nineties was alive in the Googleplex.

■

GOOGLE HAD ALWAYS BEEN OBSESSED with its logs, the reams of data its users provided by their billions of searches. Google's engineers used this data to improve the algorithms, but as the company was committed to "organizing the world's information," it also had a fascination with how search behavior revealed the world's obsessions in real time. Eventually, products like Google Trends and Google Zeitgeist would allow us all to peek inside the planet's collective unconscious, surfacing perennial obsessions like "sex" or "porn" but also faddish searches like "Paris Hilton" or "Justin Bieber." In the year 2000, the hot search term was "MP3." This was because, across the country, a teenager just barely into his first year of college had dreamed up a program that would break the Internet wide open just as definitively as Google's algorithms were doing.

Shawn Fanning was a member of the first true web generation, born November 22, 1980, in the working-class Boston suburb of Brockton, Massachusetts. Earlier than most people his age, Shawn became a heavy user of online chat, especially Internet Relay Chat. It was on IRC that Shawn Fanning fell deeply in with the teenage hacker crowd.

Sometime in 1997 or 1998, Shawn was invited to join the private IRC channel called woowoo, which was the main online meeting place for a hacking collective of the same name. Members of woowoo would go on to have a hand in the formation of dozens of technology companies ranging from WhatsApp to Arbor Networks, but at the time, they were just a bunch of kids trading hacks.[23] Under the pseudonymous login handle "napster," Fanning traded programs and coding advice, trying to impress the other hackers with exploits and programs he scratched together himself.

In the fall of 1998, Shawn enrolled at Boston's Northeastern University and saw that his new roommates and fellow students were obsessed with finding and trading music files known as MP3s. But finding these files was a complicated process of searching FTP (File Transfer Protocol) sites, Usenet newsgroups and other online repositories. There was also no real way for users to exchange these files easily among themselves. So, late in 1998, Shawn Fanning announced to his fellow hackers on woowoo that he was working on a program that would make finding and exchanging MP3 files a breeze.

■

FROM THE EARLIEST DAYS, people had dreamed of turning the web into a medium for music. As early as 1993, two students at UC Santa Cruz launched a website called the Internet Underground Music Archive so that artists and musicians could upload and distribute digitized recordings for others to download and listen to. This proved popular, but largely unwieldy for most users, since the size of the music files was too large for the dial-up Internet connections of the day; downloading a single song could take half a day to complete. This changed in the mid-nineties, when a new type of music file was introduced. ISO-MPEG Audio Layer-3, or MP3, was developed at the Fraunhofer Society for the Advancement of Applied Research in Germany and used audio and file compression to create music files that were much smaller in size, but without sacrificing too much in the way of sound quality.

It turns out that the human auditory system is not an instrument that scoops up all the frequencies in a given environment, like a microphone does. What we "hear" is not an accurate representation of reality, but only those sounds that the brain, over the course of millenia of evolution, has determined to be the "most important" sounds. By stripping out the unnecessary (because they were unheard) noises in a sound file, music files could be made much smaller. Most music was easily compressed and a listener was none the wiser. "That's an undergraduate project," says Karlheinz Brandenburg, the Fraunhofer researcher who is called the "father" of the MP3.[24] But the human voice was far trickier. It turned out that the key to mastering the nuances of human singing was an obscure a cappella recording of a minor hit from the 1980s, Suzanne Vega's "Tom's Diner." Brandenburg successfully tweaked the MP3's compression algorithm by listening to "I am sitting / In the morning / At the diner / On the corner . . ." over and over again, maybe 10,000 times, before he got it right. "To get it to the level that it's really perfect, or near-perfect, for everything," says Brandenburg, "*that* was work."[25]

The resulting files were small enough to be useful in a low-bandwidth era, but MP3 technology further benefited from another technological leap that was occurring at just the same time: computer storage was exploding. The web had been born in an era when the average computer hard drive was still measured in megabytes. The first gigabyte hard drives only became commercially available in the mid-1990s,[26] and by 1999, CNN was trumpeting the arrival of 5GB, even 10GB, hard drives.[27] That amount of storage might seem woefully small for even a smartphone these days, but in the late 1990s, it was a massive amount, more than enough to store not just numerous songs, but entire albums worth of MP3s.

The media was there, the storage was there, and just as serendipitously, the ability to play this media arrived on the scene as well.

In 1997, a nineteen-year-old college dropout named Justin Frankel released a software program called Winamp, which allowed users to easily organize and play MP3s on computers. Winamp was downloaded by more than 25 million eager MP3 devotees, and Nullsoft, Winamp's parent company (which Frankel had formed with the Internet Underground Music Archive's Rob Lord), was sold to AOL in 1999 for around $100 million.[28]

In a way, Shawn Fanning was trying to solve the final piece of this puzzle: a search engine for MP3s. But since most MP3s were sitting on individual users' computers, he needed to find a way to search other people's hard drives, not public webpages. That way, if you wanted to find a particular song, you could simply figure out who had it on their computer and get it directly from them. You would share the songs on your hard drive as well, thereby keeping the karmic cycle going. Fanning's MP3 search program would be networking in its purest form; it would be a literal peer-to-peer exchange.

"It felt like this way of sharing media between people could be used for sharing anything," Fanning would say later. "It also felt like this whole model for sharing media was superior to, like, going and buying an album. . . . Basically to have access to the entire universe of recorded music. . . . In every way it seemed like a better system."[29]

■

IN A FEW SHORT WEEKS, Fanning coded up a rough version of a program, which he named after his nom-de-hacker, Napster. As was the custom, he turned to the other hackers in woowoo for tips and advice. Among those other woowoo users who began chipping in to contribute to the program was a slightly older, slightly more sophisticated coder named Jordan Ritter (woowoo handle: "nocarrier") and a less technically savvy but more ambitious woowoo hanger-on named Sean Parker (nickname: "nob"). Ritter would eventually take over the sophisticated back end of the Napster system, developing the complicated server connections, search algorithms and networking details that would allow users to search each other's computers and download MP3s directly among themselves. And as for Parker's contribution? Well, Sean Parker wanted to turn Napster into a business.

Despite the populist image it cultivated later, Napster was conceived of as a business from day one. The Napster phenomenon was covered in the press as some sort of grass-roots movement that bubbled up out of nowhere, largely because that was the image Napster, the company, later fed to the press. But the truth is that long before Napster was a multimillion-user phenomenon—before Napster even had users numbering in the tens of thousands—the idea was to turn Napster into a billion-dollar company. This inclination was partially a result of the time Napster was born into; 1998 into 1999, when Napster was being developed, was the height of the dot-com mania. But it was also because the brilliance of the Napster idea was immediately obvious to everyone involved: it was an entirely new way to distribute media. Imagine being able to search and instantly find any song in existence. And then imagine the instant gratification of being able to download those songs and play them right away. Oh, and by the way: all those songs were completely, 100% free of charge, because you were getting them, not from a record store, but from some other, unknown Internet user.

Napster was seeking to raise money from investors not long after it left the friendly confines of the woowoo IRC channel. This was thanks to the precocious Parker, who took it upon himself to raise money for the project, running through a chain of connections that eventually landed Napster a $250,000 investment from a California angel investor on Labor Day 1999. By the fall of 1999, Shawn Fanning, Sean Parker, Jordan Ritter and another woowoo regular, Ali Aydar (IRC handle: "mars") were out in California turning Napster into a real startup.

Napster was like a supernova that exploded across the tech, media and cultural landscape just as the dot-com bubble burst in the year 2000. The grand-slam idea that everyone saw in Napster's technology proved itself out spectacularly. By the spring of 2000, less than a year after launching, Napster had more than 10 million users.[30] By the end of 2000, Napster could claim more users than even mighty AOL: around 40 million. And instead of taking more than a decade and billions of dollars to do so, Napster had attracted that many users on the backs of half a dozen barely postpubescent hackers and about $400,000 worth of hardware.[31]

Napster owed its success to all those college kids with their gigabyte hard drives and broadband dorm room Internet connections. By the spring college semester of 2000, an estimated 73% of college students were using Napster regularly.[32] On some campuses, Napster was consuming nearly 85% of available bandwidth.[33] When various institutions began enforcing Napster bans, students nearly rioted. For a long time, Napster was in *The Guinness Book of World Records* as the fastest-growing service of all time.[34] At points early on in its development, Napster's user numbers were growing 35% *a day*.[35]

But if Napster was a supernova, it was also *the* star-crossed startup of the Internet Era. Even nearly twenty years on, it's hard to imagine how Napster could have ever succeeded. And that's before taking into account all of the self-inflicted wounds the company visited upon itself.

A series of management regimes were recruited to try to build Napster into a proper company, but as Jordan Ritter has said of the quality of leadership Napster was able to bring in, "You would think the truly fastest growing Internet startup in the world would attract the best people. But it did not. It attracted the worst people."[36] Napster was not able to attract the best investors either. Unlike Google, which was raising money at almost the exact same time, Napster never landed a deal with the VC blue chips like Kleiner Perkins (though Kleiner took a hard look before passing).

It turned out that Napster's biggest problem was what it *actually did*: allow users to exchange copyrighted songs for free. It allowed people to pirate music. It was hard to argue that this was not, at least in some way, illegal, and that was what scared off the blue-chip investors and big-name management types. Napster would argue vehemently that it was merely a middleman; a technology that allowed users to connect; in some ways it was no different than an ISP like AOL or a web service like Yahoo. People could—and did—exchange copyrighted material on AOL all the time, and no one argued that AOL was illegal. To this day, Napster insiders like

Jordan Ritter believe that there was a sound legal loophole for Napster.[37] In an age of computer networks, how did it make sense to blame a technology itself for how its users employed that technology? Ever since the advent of the CD, music was nothing more than ones and zeros, digital lines of computer code. When you bought a physical album, you had always been allowed to give it to your friend or make them a mix tape from it. Because you could now do the same thing digitally, because you could now store your entire music collection on your hard drive instead of on shelves—how did that suddenly make it wrong to do with your music what you wanted?

Nonetheless, the legal aspects of what was happening on Napster's network were new and untested by precedent. Everyone knew that it was only a matter of time before Napster wound up in court, and sure enough, on December 6, 1999, the Recording Industry Association of America filed a lawsuit against Napster in San Francisco's U.S. District Court. Napster was not even six months old.

This is another point that's widely misunderstood about the Napster story. The lawsuits and the media publicity that came with them helped *create* the Napster sensation. It was almost a textbook example of the Streisand Effect, the phenomenon (as Wikipedia describes it) whereby an attempt to hide, remove or censor a piece of information has the unintended consequence of publicizing the information more widely. Before the lawsuit, there were maybe 50,000 users on Napster; a month after the lawsuit, that number had tripled to 150,000.[38] By the summer of 2000, there were more than 20 million.[39] The phenomenon of Napster, this seemingly organic impulse that suddenly inspired millions of everyday people to skirt copyright laws and social conventions and begin exchanging music freely with one another—it was largely inspired by the publicity surrounding Napster's legal battles.

Napster played up the publicity for all it was worth. It cast itself simultaneously as (1) the little guy getting beat up on by greedy corporations, (2) the cutting-edge technology company that the dinosaurs of old media were threatened by and (3) the champion of everyday users who just wanted to consume their music the way they wanted. Napster quietly encouraged the campus protests when the RIAA pressured colleges to block Napster from their networks. As would later come out during litigation, Napster even paid some musicians to publicly support the service, encouraging them to laud Napster as a foil to the rapacious record industry. And when the vociferously anti-Napster band Metallica showed up at Napster's offices to deliver a list of more than 300,000 Napster users it claimed were pirating the band's tracks online, Napster organized a "spontaneous" same-day counterprotest to ensure that the event made front-page headlines. "Fuck you, Lars, it's our music too!" protesters shouted at Metallica's Lars Ulrich as he delivered the list of usernames.[40]

Napster also played up the by now well-worn angle of a young company founded by a bunch of kids who just wanted to change the world. Shawn Fanning and Sean Parker were paraded regularly on MTV and other television outlets. Napster made the cover of magazines from *Rolling Stone* to *Time*. Shawn Fanning introduced Britney Spears at the 2000 MTV Video Music Awards and hobnobbed publicly with famous artists such as Billy Corgan and Courtney Love. Fanning even testified before Congress alongside Metallica's Ulrich.

But the bottom line was that Napster users *were* pirating copyrighted songs, and it was this simple fact that Napster couldn't escape. Napster hired lawyer David Boies, fresh off his victory over Microsoft, to argue that Napster didn't have any control over what its users did, that its servers didn't touch, much less store, any of the copyrighted material, that it was no more liable for crimes committed because of its technology than the phone company was for allowing users to dial in to Napster in the first place. But none of it mattered in the end, because the courts decided that Napster *knew*; it knew what its users were up to, and that made all the difference in the world.

Napster was ultimately done in by internal documents that were uncovered during the RIAA trial. In a key email exchange between Shawn Fanning and Sean Parker (who was, ostensibly, the strategic visionary of the early Napster), Parker wrote about the need for Napster users to protect their anonymity: "Users will understand that they are improving their experience by providing information about their tastes without linking that information to a name or address or other sensitive data that might endanger them (*especially since they are exchanging pirated music*)."[41] The emphasis on that last statement is mine, but at trial, the RIAA stressed that section as well. In her initial ruling against Napster, the judge in the case, Marilyn Hall Patel, ruled that the evidence "overwhelmingly establishes that the defendant had actual or, at the very least, constructive knowledge" that users were using Napster to pirate copyrighted music.[42] Napster briefly got relief on appeal, but ultimately, rulings came down that said the company either had to put a system in place that blocked copyrighted material on its network, or else it had to shut down the entire network. Fanning and the other Napster engineers tried gamely to implement algorithms to do just that, and they succeeded in blocking 98% to 99% of the offending material. But the judge was ultimately not satisfied unless the percentage of blocked material reached 100%, and Napster was never quite able to achieve that. When all legal options were exhausted, Napster filed for bankruptcy on May 14, 2002, and fired all seventy employees, including Shawn Fanning, who had stayed with his brainchild until the bitter end (Jordan Ritter had left in October of 2000, and Sean Parker had been quietly shown the door after his damning emails had come to light).[43]

Napster was perhaps the victim of its own naïve faith in technology. Did Napster know that people were largely using its technology for pirating music? "Yeah we knew," Napster engineer Ali Aydar would say years later. "But we also knew that this thing called the Internet existed. And it was new. And as it evolved, these things were going to start to happen. And things were going to have to change. And the way in which the world worked was going to have to change."[44] The hope was that if the majority of the music-buying public could be converted to this new way of consuming music—of downloading, of storing songs on your hard drive, of every song in the world being available at your fingertips—that Napster could then cut a deal with the record companies, something along the lines of "Hey, all your customers are now on our platform. Let us help you reach them, in a mutually beneficial, profitable way." In internal strategy documents drawn up by Parker, this was laid out explicitly: "We use the hook of our existing approach to grow our user base, and then use this user base coupled with advanced technology to leverage the record companies into a deal."[45]

Surely the record companies would see that digital distribution was more efficient. They would see that Napster could help people discover new artists and promote existing ones by creating a central hub. In retrospect, there is no shortage of people, even inside the music industry, who imagine how different the world would be if it had worked out that way—if the music companies had partnered with Napster and accepted the inevitability of technology. "Something like thirty million–plus music fans were in one spot online," says Jeff Kwatinetz, a former representative of music artists ranging from Linkin Park to Mandy Moore and Ice Cube. "At the time, the idea of all the music you would want for $15 a month was an appealing thing and studies showed most users would have paid it."[46] Napster could have been *the* portal for all of music, a Yahoo of music, a Google of music, maybe even a Facebook of music.

Of course, a less polite word for "leverage" is "extortion." Perhaps Napster's biggest misstep was trying to leverage the record companies into a deal, given that the music business has always been known as one of the most notoriously cutthroat and aggressive in the world. This was an industry with quite literal mob ties throughout much of its existence. Napster simply picked a fight with the wrong adversary. The music industry was never interested in a deal. The music industry was only ever interested in suing Napster dead.

The RIAA would follow up its victory over Napster by attempting to sue other digital technologies out of existence, and even, eventually, suing music consumers themselves—tens of thousands of them, in fact. Of course, all this did nothing to halt the advance of file-sharing technology. In Napster's wake, first came Gnutella, from Justin Frankel, who had created Winamp. Gnutella spawned a whole ecosystem of next-generation file-sharing networks like LimeWire, BearShare, Morpheus and many more. A few years later, in 2003, a twenty-five-year-old coder named Bram Cohen released the BitTorrent protocol, which took file sharing to new frontiers like movies, TV shows, and video games.

If Napster had been naïve to think it could have done a deal with the record companies, then the record companies were certainly naïve to think destroying Napster would somehow make the threat of digital technology go away. But, as has been endlessly discussed and is widely understood, the music industry was caught in a classic innovator's dilemma, tied to a highly lucrative business model it was loath to give up, even in the face of an existential threat presented by new technology. Everyone knew the music industry had gotten filthy stinking rich on the back of the compact disc in the 1980s and 1990s. Having convinced all of us to repurchase our record collections in digital form, the music industry went from selling 800,000 CDs in 1983 to 288 million in 1990 and nearly a billion in the year 2000.[47] Unlike with most digital technologies (the price of which almost always declines over time), the price of the average CD seemed only to inch upward every year, approaching nearly $20 a disc by the turn of the century.

But even this analysis—the record companies were wedded to the cash cow of the CD—doesn't quite get at the truth behind the revolution that Napster began. Napster was the first signal that the web had changed consumer behavior in a fundamental way. Today, we live in a world where consumers not only expect, but demand, infinite selection and instant gratification. Amazon had first introduced the concept of infinite selection, and now Napster was training an entire generation to require the instant gratification. Shawn Fanning had been right from the very beginning: digital really was a better way to distribute music. Computers (at least, the gadgetry computers would evolve into) would turn out to be pretty damn good music consumption machines.

Advertising might have been the first industry the web disrupted, but Madison Avenue adapted to the change, quickly following our attention spans and our eyeballs as they drifted online. The record companies, in contrast, refused to budge as the habits and preferences of music consumers changed. It was never piracy that was the problem for the music industry (at least, not entirely). But rather, it was the stubborn refusal to adapt to a revolution in consumer expectations that has, at its root, truly bedeviled the record companies, and the television companies and the movie companies, and on and on and on over the course of the Internet Era.

Infinite selection. Instant gratification. On any device. When it comes to digital disruption of media, it is almost never about free content or piracy, not at the core. It is *always* about giving people what they want, when they want it, how they want it. Napster seemed to understand this intuitively, even if its execution on this insight was bungled. In early interviews where Shawn and Sean were trotted before the media to explain what Napster was trying to do, Sean Parker would say things that, in retrospect, were completely dead-on. "Music will be ubiquitous and we believe you'll be able to get it on your cell phone, you'll be able to get it on your stereo, you'll be able to get it on whatever the device of the future is. And . . . I think people are willing to pay for convenience."[48] The Internet and the web and Google had already made information ubiquitous. Napster was the first company to prove that, in the future, media would be ubiquitous as well.

■

EVERYONE TENDS TO FOCUS on the Napster trial as the pivot point in the history of modern technology versus traditional media. But there was another trial, from around the same time, that would ultimately have a larger impact on how we consume media in the digital era. In September 1998, a small company called Diamond Multimedia released one of the first portable MP3 players, the Rio PMP300. The PMP300 had only 32 megabytes of storage, so it could only hold about 30 minutes of music—half an album or so, at decent sound quality; a whole album and a couple extra songs if you didn't mind compressing everything to a level of barely tolerable sound quality.[49] About a year before it sued Napster, the RIAA sued Diamond Multimedia. Before it had even heard of Napster, the record industry knew it didn't want MP3 as a technology to catch on. But while Napster was eventually defeated, the RIAA *lost* the Diamond Multimedia case. The Rio PMP300 went on to become the first commercially successful portable MP3 player.

As the author Stephen Witt has noted in his book *How Music Got Free: A Story of Obsession and Invention*, from the perspective of history, the music industry won the wrong lawsuit.[50]

12

RIP. MIX. BURN.

The iPod, iTunes and Netflix

At the turn of the century, Apple was not anywhere close to the technology behemoth it is today. In fact, Apple was dangerously close to bankruptcy before its original cofounder Steve Jobs came back to Apple and saved the company. Because Apple was such a niche player in its main market, personal computers, it didn't have much of a role in the first part of the Internet Era. In the late 1990s, Apple's share of the computer market in the United States fell below 3%.[1] The dot-com world was a Windows world, and so Apple simply wasn't a company Internet players paid much attention to.

It's legendary now how Jobs returned to Apple, ruthlessly streamlined its product lines by killing extraneous projects, brought in a young executive named Tim Cook to turn Apple's supply-chain and manufacturing processes into the envy of the technology world, and embraced a then-obscure industrial designer named Jonathan "Jony" Ive to churn out innovative and beautifully crafted computers that stood out from the dull beige boxes produced by the likes of Dell and Compaq. The product that would change Apple's fortunes was the iMac, the Ive-designed translucent and colorful computer that debuted in 1998 and would become Apple's biggest hit in a decade. The "i" in iMac was meant to suggest an innovative, but also individualized, device—a return to the "personal" in personal computing. But the "i" was also meant to suggest "Internet."[2] In a time of near-total Windows domination, Jobs and Apple struck on the idea of rebranding their Macs as machines uniquely designed for the Internet Era.

Jobs wanted Macs to become the "digital hub" in a forthcoming media utopia that would include then-novel gadgets like DVD players, digital cameras, digital camcorders, personal digital assistants and more. The Mac would be the central machine that would help manage and empower all the other gadgetry. To this end, Apple began releasing a whole suite of Apple-produced (and Macintosh-exclusive) software applications to make the digital hub a reality, including iDVD, iPhoto, iWeb and GarageBand.

But, thanks to Napster, digital music was all the rage. The next logical step in the digital-hub strategy would be to create a killer application for digital music, for MP3 files. One of the more popular Mac applications for MP3s was a Winamp-like digital jukebox called SoundJam, which was developed by two former Apple engineers. In March of 2000, Apple purchased SoundJam and turned it into an application called iTunes, which was unveiled at the Macworld trade conference on January 9, 2001.[3] iTunes became the flagship of Apple's digital hub. Combined with the new CD burners that Apple began to ship in its computers, users were soon encouraged to "Rip. Mix. Burn." In other words, digitize your CD collection, create playlists and mix songs like you did on programs like Winamp, and then burn CDs of your choosing.

Steve Jobs intuited that music was the key spoke in the wheel that was the digital hub. But on the outer edge of that wheel were the devices that the hub was supposed to manage, and when Steve Jobs looked around at the digital devices for music, the MP3 players produced by companies like Diamond Multimedia, he felt that they, in his words, "truly sucked."[4]

The first MP3 player to be widely commercially available to consumers was the MPMan released by a South Korean company named Saehan Information Systems in 1997.[5] The earliest MP3 players were developed by smaller companies like Saehan and Diamond Multimedia because of the questionable legal status of the MP3. After the

RIAA lost its case against Diamond, a flood of other devices followed from an array of companies. But it was this fragmented market that was also the main reason the early MP3 players "sucked." The smaller companies didn't have much experience with hardware or software design. It was difficult to get music onto the devices, to manage the music files, and they couldn't store many songs.

Well—Apple was the best in the world at user interface design.

But—Apple wasn't really in the gadget business. It was in the computer business.

Although—come to think of it, what *were* MP3 players but tiny, single-purpose portable computers?

In late 2000, an Apple executive named Jon Rubinstein made a routine trip to Japan to visit electronics supplier Toshiba. In his meeting with Toshiba's engineers, he was told about a new, incredibly small, 1.8-inch hard drive the company had developed that could hold up to five gigabytes of data. Toshiba wasn't sure what it could be used for. Laptops, obviously. Or maybe digital cameras? Rubinstein knew *exactly* what it could be used for. Toshiba's hard drive was the size of a silver dollar but had the capacity to store about 1,000 MP3s. If Apple married this hard drive to its elegant hardware and software design prowess, it could design an MP3 player that blew the others out of the water. Jobs authorized Rubinstein to buy all the 1.8-inch hard drives he could get his hands on.

The device that would eventually be called the iPod was the result of a crash development program that took place over less than a year. Rubinstein led the iPod development team along with Tony Fadell, a gadget whiz who had previously worked at Philips Electronics. Apple did indeed bring its unique design magic to the project, under Jobs's famously exacting tutelage. In April of 2001, the iPod team presented their prototypes to Jobs in person. Fadell, who had only recently joined the company, had been warned by veteran Apple executives that Jobs tended to reject early ideas, no matter their merit. So, Fadell presented other concepts first, saving his favorite prototype for the end of his presentation: a rectangular device about the size of a pack of cigarettes—small enough to fit in pants pockets. Another Apple executive, Phil Schiller, demoed the iPod's key user-interface breakthrough. With so many songs to choose from, selecting something to listen to could get tedious. "You can't press a button hundreds of times," Schiller said.[6] Instead, the iPod would feature a wheel that allowed you to scroll quickly through your list of files.

"That's it!" Jobs shouted.[7]

In later meetings, Jony Ive would contribute to the device's iconic aesthetic. "Right from the very first time we were thinking about the product, we'd seen this as stainless steel and white," Ive said later. "It is just so . . . so brutally simple. It's not just a color. Supposedly neutral—but just an unmistakable, shocking neutral."[8] Even the headphones would be white. Well, not exactly white, because Steve Jobs hated pure white. So, the headphones were technically a shade named moon gray, which was so light as to appear white.[9]

Jobs would later remember the development of the iPod as a series of serendipities. "We suddenly were looking at one another and saying, 'This is going to be so cool,'" Jobs told his biographer, Walter Isaacson. "We knew how cool it was, because we knew how badly we each wanted one personally."[10] The device was at once a departure for Apple—a leap into the consumer electronics space—and the purest expression of the digital hub philosophy. Combined with the iTunes software app, users would have full control over their digital music collections.

Steve Jobs unveiled the iPod on October 23, 2001. With his showman's panache, he emphasized the device's enormous capacity—"You can take your whole music library with you"—combined with its extreme portability—"I happen to have one right here in my pocket."[11] The price was on the high side: $399, and it was only available to users of Apple's Mac computers. Nonetheless, Apple sold 150,000 iPods in the first quarter they went on sale.[12] Users could now organize their music using Apple's carefully designed iTunes software, and enjoy their music on Apple's beautiful iPod. But there was still one part of the equation that was missing: how were you supposed to acquire the MP3 files to fill your iPod and your iTunes library? Well, you would rip them from your existing CD collection or download them from the file-sharing sites, because the record companies still weren't embracing the MP3 format.

Napster and the flood of technologies that followed in its wake hadn't impacted the music industry right away. In the first half of 2000, while the RIAA was in the midst of shutting Napster down, music sales were actually up 8%.[13] But the next year, the impact of file sharing appeared. In 2001, the number of CDs burned worldwide matched the number of CDs sold in retail stores. Consumers didn't need to buy prerecorded CDs when they could rip, mix and burn their own. By 2003, it was estimated that 2 billion music files were being exchanged *every month*. More than 57 million people were sharing music files in the United States alone.[14]

By 2002, music revenue came in at $12.9 billion, down 13.7% from its $14.6 billion peak in 1999.[15] The industry was increasingly desperate, looking for any solutions that would stop the bleeding. They had tried to launch their own online music download sites (with stringent digital rights management software to prevent the proliferation of illegal copies, naturally), but these initiatives generated little interest from the public at large and collapsed under the weight of industry infighting and competing strategies.

Into this crisis stepped Steve Jobs. With iTunes and the iPod, Apple had crafted an end-to-end software and hardware experience that was increasingly popular but also carefully managed. All it needed was a commerce element so that a fully digital musical ecosystem could be realized. It was clear that this was what users wanted, and if the record companies had been afraid to embrace the digital future for fear of piracy and file sharing, well, Apple was willing to tackle that problem as well. Jobs directed Apple's engineers to create a digital rights management system called FairPlay that carefully limited the devices music files could be played on. He did this not because he liked copy protection schemes (he didn't; he felt DRM needlessly complicated the user experience), but because he knew that was the way to get music executives to the negotiating table.

In early 2002, Jobs began approaching executives at the five major record companies with a proposal to create an iTunes store. If the record companies would license him the right to sell their catalogs as digital downloads, Apple would make sure that users would enjoy their music inside Apple's end-to-end ecosystem in a controlled and, crucially, legal way. Still the record companies balked. Jobs was insisting that they learn *the* crucial lesson of Napster: consumers wanted unlimited selection and freedom of choice. People were filling their hard drives—and now their iPods—with their favorite songs, not necessarily their favorite albums. Jobs wanted to sell individual songs on the iTunes store, and this was what the record companies couldn't abide because they were still wedded to the physical album.

Jobs was convinced that ease of use and customer choice were key to competing with the lure of the free. Making it dead simple for people to get what they wanted would make piracy seem like a hassle in comparison. Sure, people could buy albums on the iTunes store if they preferred, but at a more reasonable price that took into account the cost savings of digital distribution. Part of that was offering individual songs at 99 cents apiece. That way, it felt like an impulse to buy music, almost an afterthought. The revenue arrangement Jobs offered—the record companies would take two-thirds of every sale and Apple one-third—was actually a better deal than the record companies got from physical retailers. And yet it was this insistence on breaking up the album bundle that almost kept the iTunes store from happening. "I've never spent so much of my time trying to convince people to do the right thing for themselves," Jobs would say later of his negotiations with the industry.[16]

In the end, Jobs was able to get the record companies—all five of them—to sell their music in the iTunes store for three important reasons. First, Steve Jobs was a rock star in his own right, an icon. As the owner of the Pixar movie studio, he was a bona fide Hollywood mogul, and to an industry of tastemakers, his celebrity had impact. Second, the music industry was desperate. Nearly three years into the era of file sharing, they had been unable to come up with a legal way to download digital music that people wanted to use. They were looking like idiots; worse, they were looking like obstructionists, unwilling to deliver what their customers were demanding. But third, they were willing to experiment with Apple because, at the time, Apple was an insignificant player. Both iTunes and the iPod were, at this point, available only to Mac users. "We used our small market share to our advantage by arguing that if the store turned out to be destructive it wouldn't destroy the entire universe," Jobs said later.[17]

Apple announced the iTunes Store on April 28, 2003, and in no time Jobs's notion that ease of use and user freedom could give piracy a run for its money proved prescient. The iTunes Store sold a million songs in just six days.[18] A year later, Apple would announce it had sold 100 million songs. Less than a year after that, 1 billion.[19]

But Steve Jobs, of course, had a legendary stubborn streak of his own. Jobs had always conceived of the iPod as a way to sell more Apple computers. He was still married to the idea of the Mac as the digital hub, so he was reluctant to bring iTunes to Windows machines (and thus, the majority of computer users). "It was a really big argument for months," Jobs recalled, "me against everyone else."[20] Jobs declared that Apple would do a Windows version of iTunes "over my dead body." Only after Apple executives showed him business studies that proved Mac sales would be unaffected did Jobs capitulate, saying, "Screw it! I'm sick of listening to you assholes. Go do whatever the hell you want."[21]

The contracts with the record companies were renegotiated and iTunes for Windows was announced in October 2003. Opening up the iPod and iTunes to Windows was the turning point that set Apple on the path to becoming the biggest, most profitable company in the world. Around the time of iTunes for Windows, Apple sold its 1 millionth iPod. That next holiday season of 2003, it would sell nearly three-quarters of a million more. A year later, in the holiday season of 2004, it sold 4.5 million.[22] By mid-2006, Apple had sold 58 million iPods in total, and the iPod-iTunes business combined contributed 61% of Apple's total revenue.[23] Apple was no longer "just" a computer company.

But then, the iPod was the first device to prove that computers were no longer just computers. In the Internet Era, you could put a computer into any piece of consumer electronics and suddenly it became something more. Twenty years before the iPod, the Sony Walkman sold 340 million units by proving that music could be both portable and personal.[24] But the iPod brought "personal" computing out into the larger world in a way that had never happened before. iPods, with their soon-to-be-ubiquitous white(ish) earbuds, became fashion statements, calling cards of the "hip" and the "modern."

The iPod returned Apple to a prominent place in the technology industry. Other companies, especially Microsoft, with its Zune MP3 player and accompanying music store, attempted to usurp Apple's dominance in digital music. But (irony of ironies) the iTunes software platform gave Apple a near-monopoly of the MP3 player and digital music download markets. It was (ironically) a software platform that Microsoft couldn't penetrate. By 2007, Apple's iTunes controlled 70% of legal digital music sales.[25] The iPod claimed a similar share of the MP3 player market.

Apple taught the world what it meant to be a consumer electronics company in the Internet Era. This was best exemplified by the brief, glorious life of the iPod Mini. The first smaller, cheaper iPod (also, the first with fashionable colors beyond white), the Mini was the model that really caused iPod sales to take off. It was the best-selling model of the iPod in its time—by far. Most other companies would have milked a cash cow product like that for as long as they could. Not Apple. Less than two years after releasing the Mini, Apple replaced it with the iPod Nano, which switched out the tiny hard drives in favor of superior Flash technology for storage, thereby allowing iPods to get thinner and more portable still. Apple showed a willingness to eat its young in order to stay on the cutting edge; to out-innovate itself before others ever had the chance.

■

NAPSTER HAD SHOWN that computers and the Internet were made for music, just as Shawn Fanning had said all along. Apple was the first company to bet its entire business on the bonanza that the new Internet paradigm made possible. But what Apple *didn't* do was save the music industry.

All through the height of the iTunes Store's popularity, the music industry's revenues continued to collapse, from an inflation-adjusted height of $21 billion in global sales in 1999 to just under $7 billion in 2015.[26] Sales of digital music didn't surpass physical music until around 2011; as recently as 2017, 22% of all music sales were still in physical formats.[27] Business models, even when they're dying, can stubbornly cling to life, right until the very end (just ask the newspaper industry, which still makes the bulk of its money delivering news on dead trees via trucks).

The record companies had been right all along: selling songs individually wasn't as lucrative as selling entire albums. Selling two or three good songs at 99 cents apiece could never hold a candle to selling one good song and eleven crap songs for $17.99. Music journalist and *Rolling Stone* editor Steve Knopper points to the Baha Men as an almost quintessential one-hit-wonder musical act. He notes that in the year 2000, their song "Who Let the Dogs Out?" was a sensation, and fans bought 4 million copies of the Baha Men album of the same name—most likely *just* to get that one song. Just three years later, the hot song was "Stacy's Mom" by Fountains of Wayne.[28] The album with "Stacy's Mom" on it only sold 400,000 copies. "Stacy's Mom" *did* sell 520,000 downloads as a digital single on platforms like the iTunes Store. But what do you think the record company would have been happier with? A hit single moving 4 million albums at around $17 a piece, or just 400,000 albums and half a million digital singles sold at 99 cents apiece?

Steve Jobs had leveraged the music industry's crisis over piracy to destroy the business model of the album. He had done so out of selfish interests, but that doesn't change the fact that his actions actually served the interests of consumers. The lesson of commerce in the Internet Era—from Amazon through Napster through the iTunes store—has been that consumer habits and expectations have changed radically. The general public has intuited that the Internet and digital technology enable a world of unlimited selection and instant gratification. If your business model stands in the way of that, well, consumers will just go around you. It's a lesson that the music industry failed to learn from Napster, and it's a lesson that media companies are having to re-confront again and again, even down to the present day.

▪

GIVING CONSUMERS WHAT THEY WANT was the key to Netflix's success as well. The official tale of Netflix's birth is that CEO Reed Hastings got a $40 late fee for not returning a copy of *Apollo 13* in time. Incensed, he founded a DVD-rental company that wouldn't treat its own customers so shabbily. Like eBay's Pez Dispenser, however, this late-fee-as-eureka-moment is merely a PR-friendly origin story dreamed up retroactively. Netflix was actually the brainchild of Marc Randolph, who had been an employee at one of Hastings's previous companies, Pure Atria. When Pure Atria was sold and both Hastings and Randolph were between engagements, Hastings agreed to fund Randolph's brainstorm, which was a scheme cribbed directly from Amazon: instead of Earth's biggest bookstore, Earth's biggest video rental store.

Launched on April 14, 1998, Netflix originally benefited from a secular media shift: the transition from VHS to DVD. At launch, Netflix could only boast a library of a little over 500 titles.[29] But these were nearly all the DVDs in existence at the time. In 1999, only 5% of households owned a DVD player. But that percentage more than doubled by 2000, more than doubled again to 37% by 2002, and climbed further to 65% by 2004.[30] Nearly every DVD player shipped with coupons for free Netflix rentals right in the box. Netflix gave you something to do with your shiny new movie machine. The DVD became the fastest adopted consumer electronics technology in history, and the number of DVD titles Netflix could provide exploded.

Netflix also benefited from being essentially the only game in town for a while. The incumbent rental behemoths—Blockbuster, Hollywood Video and Movie Gallery—were reluctant to embrace the new format. They had been burned by the earlier LaserDisc technology, which had only proven popular to a niche audience. When, in the summer of 2000, Netflix even offered to sell itself to Blockbuster for about $50 million with the express idea that Netflix would become the DVD channel for Blockbuster, thereby saving it from the costly transition of its inventory from VHS, Blockbuster said no.[31] It still didn't believe DVDs would catch on.

Netflix was originally launched as a hybrid service. You could rent DVDs à la carte for $4 a piece plus $2 for shipping the disc to you in those little red envelopes.[32] But you could also buy the DVDs to own, just as a dot-com named Reel.com was offering at the time. The problem was, neither option proved very profitable; as other ecommerce companies had learned, for such small-dollar items, the shipping costs really ate into margins. When Reed Hastings moved from being merely Netflix's biggest investor and advisor to becoming its full-time CEO in 1999, he discontinued the retail sales, and the company began experimenting with different rental schemes in an effort to find profitability. The model Netflix settled on in September of 1999 was originally called the "Marquee" plan. Subscribers paid $15.95 per month for the privilege of renting four (later, just three) movies per month. If you finished one movie, you simply mailed it back and Netflix would send you another. There were no other fees—especially no dreaded late fees. Hastings called the new program a "near DVD-on-Demand service."[33] Netflix's rental volume increased by 300% in just three months.

Even though it stumbled upon the strategy accidentally, Netflix would cleverly use the "no late fees" mantra as a way to position itself as the populist champion of the consumer. "Movie renters are fed up with due dates and late fees," Hastings would tell the press. "With no due dates, our customers can stock up on rental movies and always keep a few on the top of their television, ready for impulse viewing."[34] Certainly, fining your customers for using your service is never a popular business model, but late fees accounted for as much as 13.4% of revenues for a rental chain like Blockbuster.

Eliminating the late fee made for friendly headlines, but it was not what made Netflix take off.[35] What mattered was that Netflix too had learned the key lesson of retailing in the Internet Era: unlimited selection, (near-) instant gratification. Whereas a typical brick-and-mortar video rental store carried 3,000 titles, Netflix carried tens of thousands.[36] With Netflix, you could almost always get a movie you wanted to see. Compare that to the experience at a typical Blockbuster, where limited inventory meant one in five shoppers went home empty-handed. New releases were often rented out by other customers, so typical Blockbuster customers had to visit a store five consecutive weekends before they could actually take home the movie they wanted.[37] Blockbuster even had an internal term for this experience: "managed dissatisfaction."

And more important, Netflix's combination of a website storefront coupled with postal delivery proved infinitely more suited to satisfying modern customers. Early on, Netflix introduced the "Queue." You could browse the site and make a list of the movies you wanted to see, much like Amazon's shopping cart. Every time you returned a DVD to Netflix, it would automatically ship out the next title in your queue. The average number of movies in a customer queue was around fifty and went a long way to endearing customers to the service by making it feel personal. "It's our biggest switching cost," Hastings would later say, a prime reason users stayed loyal.[38] Netflix became a platform to cultivate your individual cinematic tastes.

Netflix also concentrated on other advantages the web made possible. Copying Amazon's Recommendation Engine, as well as DVD retail competitor Reel.com's pioneering Movie Match technology, Netflix developed its CineMatch movie recommendation system. Users were prompted to check out movies they might like based on previous titles delivered from their queue. Netflix invested heavily in this technology, hiring mathematicians and computer scientists to tweak the algorithm to include recommendations based on the habits of subscribers with similar taste. Netflix's recommendation engine proved so uncannily good at predicting what you might want to watch that, eventually, nearly 70% of the movies that users chose for their queues were recommended by the algorithms.[39] This was convenient for Netflix because it allowed for greater inventory and cost controls than a brick-and-mortar store could realize. Whereas nearly three-fourths of total rentals at a Blockbuster were for new releases, at Netflix, seven out of ten DVDs rented by subscribers were titles with release dates older than thirteen weeks.[40]

■

AS HAD HAPPENED WHEN Amazon first began to challenge the big-box retailers, most people assumed that once Blockbuster turned its attention to online video rentals, Netflix would be crushed. In 2002, Blockbuster spokespeople were already dismissing online rental services as "serving a niche market."[41] But soon enough, the entire rental industry began to feel the pressure of online competition. By 2002, Netflix had attracted 750,000 subscribers, which, while only representing 2% of the video rental market, nevertheless caused sales to slip at the rental chains, especially in areas where Netflix was popular.[42] Perhaps two years too late, Blockbuster Online was rolled out as a direct web-and-postal-delivery competitor to Netflix in 2004. As Barnes & Noble had, Blockbuster tried to leverage its physical footprint. You could return movies by mail or at your local store. It also unveiled a "no late fees" program in 2005, which shut Netflix up about that feature, but had the simultaneous effect of costing the company about $600 million in lost revenue.[43]

Unlike with Amazon and the booksellers, once Netflix began to eat into the video rental market, the decline of the retail rental industry came quickly. At its peak, Blockbuster had more than 10,000 stores and 50 million members.[44] At one time, more people had Blockbuster cards than American Express cards.[45] But by 2010, Blockbuster had only 25 million customers and 4,000 remaining stores.[46] That same year, Netflix announced that it had attracted 13 million members and was mailing 2 million discs daily in the United States.[47] Blockbuster filed for bankruptcy protection on September 23, 2010.

Store closings, layoffs and bankruptcies are often the markers we use to measure the disruptive effects of ecommerce. There were nearly 25,000 individual video rental store locations at the industry's height.[48] At one point, 60,000 employees wore the blue shirts of Blockbuster.[49] Video stores were once one of the most common retail storefronts in America; there was hardly a neighborhood without one. Today, the few video rental stores left are nearly museum pieces. Netflix won not because it eliminated late fees, but, again, because it understood how consumers' expectations were changing and moved to satisfy those new expectations. Unlimited selection. Instant gratification.

And Netflix deserves credit for continuing to move in that direction even after it had conquered DVD rentals. As early as 2002, Reed Hastings was telling *Wired* magazine, "The dream 20 years from now is to have a global entertainment distribution company that provides a unique channel for film producers and studios. . . . In five to ten years, we'll have some downloadables as well as DVDs. By having both, we'll offer a full service."[50] He was talking about video on demand. About Netflix becoming a studio and producing its own content. About streaming. All delivered via the Internet.

"We named the company Netflix for a reason," Reed Hastings has said on more than one occasion. "We didn't name it DVDs-by-mail."[51]

13

A THOUSAND FLOWERS, BLOOMING

PayPal, AdWords, Google's IPO and Blogs

Netflix's successful IPO on May 23, 2002, was an early sign that the Internet was not over as a wealth-generating machine, and though it was one of the first Internet companies to go public after the bust, it was not *the* first. That was probably PayPal.

PayPal began life as Confinity, launched in July of 1999 by Peter Thiel and Max Levchin with the immodest proposal of disrupting the global financial system. From the first days of the web, people had wanted to use the Internet to create some form of ecurrency. "As far back as 1995, there were a hundred companies that used cool technologies for moving money and that were going to change the world," Thiel recalled.[1] In the midst of the bubble, there were well-funded digital money schemes like Flooz.com and Beenz.com that did not survive the nuclear winter. PayPal's crucial insight was that payments in cash could be beamed directly to your virtual person: your email address. By the end of the nineties, everyone had an email address. PayPal simply turned your email address into a virtual bank account routing number. Need to send me $10? Use PayPal to send it to my email address.

Where the virtual-bank-account-tied-to-your-email-address really found traction was among web users who were already doing a lot of virtual transactions over the web: eBay buyers and sellers. On eBay, 90% of transactions took place via check or money order.[2] Credit card merchant accounts cost hundreds, even thousands, of dollars to set up, and were designed for actual businesses. But what if you just wanted to unload your used record collection on eBay? There was no mechanism to take easy payment via credit card for the eBay hobbyist.

Enter PayPal. Sellers on eBay simply asked buyers to "PayPal" them the payment for a successful auction to their email address. PayPal would withdraw the funds from one, and forward to the other. Among the eBay community, PayPal quickly generated a strong network effect: the more sellers asked to be paid via PayPal, the more buyers were incentivized to sign up for a PayPal account, and vice versa. Just as Hotmail had advertised itself with every email sent, PayPal attracted users with every auction that was settled using its service. PayPal quickly registered 10,000 users only two months after launching, and 100,000 a mere month after that.[3]

PayPal had early competition from another company that had neighboring offices in Palo Alto. X.com was founded by a serial entrepreneur named Elon Musk, who had a vision that was just as grandiose as Thiel and Levchin's: a next-generation suite of banking and financial services that would be entirely virtual. For a while, the two competed fiercely for users, but in March of 2000, X.com and Confinity merged, eventually adopting the PayPal moniker for the combined company.

PayPal was initially completely free to use, but the service eventually charged sellers 2.9% and 30 cents per transaction—still less than credit card companies charged small merchants, and without any of the overhead or complexity. Very quickly, PayPal discovered how lucrative merely acting as a commercial middleman could be. By the fourth quarter of 2001, PayPal was profitable, thanks to facilitating payment for roughly one-fourth of eBay's total auctions. After a mere twenty-six months of operation, there were 12.8 million PayPal accounts. It had taken eBay more than four years to reach 10 million accounts.[4]

On Friday, February 15, 2002, PayPal went public and enjoyed a 55% first-day pop. The financial press, which had been so instrumental in cheerleading for the bubble, now proved downright hostile to the return of Internet IPOs. "It's an anachronism—straight out of 1999," the *New York Times* quoted a stock analyst as saying. "It's like we've kind of forgotten what got us into this situation in the first place."[5] But doubters were proved wrong a mere five months later when eBay acquired PayPal for $1.5 billion, one of the biggest acquisitions in the wake of the dot-com implosion.

PayPal showed that the web was still fertile ground for innovation, but perhaps the greater legacy for the company was how it proved to be the finishing school for an entire generation of entrepreneurs who would go on to lead the renaissance of the technology industry. Elon Musk, of course, went on to found Tesla. Peter Thiel became the first major investor in Facebook. Early PayPal employee Jeremy Stoppelman founded Yelp. Max Levchin founded slide. And PayPal alumni had a hand in founding, funding or contributing to the development of so many subsequent companies (LinkedIn, YouTube, Yammer, Palantir, and Square, just to name a few) that folks in technology often refer to a "PayPal Mafia" that runs Silicon Valley today.

■

THE SUCCESSES OF NETFLIX and PayPal were beginning to banish the ghosts of the dot-com bubble, but it wouldn't be until the undisputed star of the final wave of web startups found its footing that people were willing to believe in the Internet again.

Google was the one service that had the greatest impact after the nuclear winter, but there was one important trait that Google shared with the departed dot-coms: it wasn't making very much money. It's somewhat forgotten now, but Google existed for several years without much of a business plan. The vision Larry and Sergey had sold the venture capitalists involved a three-pronged strategy. First, Google would license its search technology to the major portals. Second, the company would sell its search technology as a product to enterprises. And third, there were some vague promises about selling ads against searches on its own website.

The young company made major progress toward the first goal by finally convincing some of the portals to use Google's results on their search pages. The first deal was struck with Netscape for its Netcenter portal, but the really big coup came when Yahoo was finally convinced to use Google for its search results (previously, a company named Inktomi had been Yahoo's search partner). The partnership with Yahoo was announced in June of 2000, and was an enormous deal for Google at the time. Part of the arrangement allowed for a "powered by Google" logo to appear on Yahoo's search pages, thereby introducing the Google brand to millions of mainstream web users. Daily searches served by Google swelled from 18 million a day before the Yahoo deal, to 60 million a day afterward.[6] In early 2001, Google would pass the 100-million-searches-per-day milestone, answering 1,000 queries a second.[7] Yahoo seemed not to mind that Google was essentially stealing its search audience because at the time it didn't feel search was a core product. It was still pursuing its portal strategy. Yahoo did, however, purchase a $10 million equity stake in its new partner, thereby tying the two companies closely together in ways that would later become important.

What Yahoo didn't know was how important the partnership would prove to be for Google's overall product. Remember that Google's algorithms improved in direct relation to the number of searches it performed and the amount of data Google's computers could hoover up. The flood of queries from Yahoo not only took Google to the next level in terms of search market share, but many Google engineers would later credit the Yahoo traffic for fine-tuning Google's search engine into its mature state. Google was essentially improving itself on the back of its biggest partner.

But the problem for Google was that the Yahoo deal simply wasn't lucrative. The fees that Yahoo coughed up were barely enough to cover the increased processing and bandwidth costs Google incurred to service the traffic. The Yahoo deal taught Google that licensing alone wouldn't be a big enough home run to build a company around—or at least, not a very big company.

The second leg of Google's original strategy was proving little better. Google produced an actual hardware device, known as the

Google Search Appliance, which was a rack-mounted box meant to be installed in corporate data centers. It was designed to provide corporations and other organizations with large amounts of data and the ability to organize, index and search that data the same way that Google did with the web. But even though Google continued to produce the Search Appliance until 2017, it never became a breakout hit.

By the end of 2000, Google was in a bit of a crisis. With monthly expenses of more than $500,000, the $25 million from Kleiner Perkins and Sequoia was starting to run low, as Google launched international versions of its website and continued to hire, taking total headcount past 100.[8] "There was a period where things were looking pretty bleak," Google board member and investor Mike Moritz admitted later. "We were burning cash, and the enterprise was rejecting us. The big licenses were very hard to negotiate."[9] And since Google had yet to earn a dime on the 70 million daily searches it was getting on its own site, by January 2001, Google's out-of-control growth was actually a problem. While the service was becoming so popular that its very name was becoming a common verb, "There was genuine concern (at the board level) about where the revenues were going to come from," says early Google investor Ram Shriram. To make matters worse, it appeared to Google's venture backers that the company's founders were reneging on their commitment to bring in a "grownup" CEO. If Page and Brin didn't recruit someone who could turn Google into a real company with real prospects to generate cash, there were rumblings that either Kleiner or Sequoia (or both) might pull out of the investment.

Of course, advertising, the third leg of Google's theoretical business model, was still an option, but in spring 2001, the existing advertising model of throwing banner ads at the top of every web page had imploded. Web advertising in general was in a deep freeze, the overall online ad market plunging to $6 billion in 2002, down from $8.2 billion in 2000. All the surviving portals were suffering because of this state of affairs.[10] In the midst of the freefall in its stock price, Google's erstwhile partner Yahoo was forced to lower its revenue guidance to Wall Street by 25% *twice* in a single quarter as the dot-coms went bankrupt and advertisers ponied up 50% less for online ads.[11]

Google had never really experimented with ads, because the company's founders were originally firmly against the idea. In their 1998 academic paper introducing Backrub/PageRank, Page and Brin had attacked the very notion of search companies relying on advertising to generate revenue because it made them "inherently biased towards the advertisers and away from the needs of the consumers."[12] In other words, ads guaranteed bad search results.

But at this very moment of crisis, a revolution in online advertising was taking place that would ultimately prove to be Google's salvation.

■

LAUNCHED AT THE TED CONFERENCE in February 1998 by entrepreneur Bill Gross, GoTo.com was conceived of as a completely new type of search engine. Instead of search results generated by spidering the web and returning pages based on an algorithm, GoTo returned results that were almost exclusively provided by sponsors. GoTo served up text ads designed to look like search results, but which were paid for by advertisers who bid for position. It was an eBay-like auction model. For any given keyword, a company could pay whatever it cost to rank first for that search term. If you wanted to show up first on a search for "flowers," you could bid, say, 10 cents a click. If someone bid 7 cents, they were listed second. Bidding a nickel might get you third place, and so on. If you wanted to go crazy and bid $1,000 a click, you could theoretically rank number one for any search term you wanted.

The idea of a "search" engine that only returned ads was extremely distasteful to most; indeed, Gross was nearly hissed off the TED stage during his presentation. But advertisers loved the idea, and signed up in droves because they quickly intuited that Bill Gross had stumbled upon one of the greatest advertising models in the history of the world. Paid search represents a uniquely powerful nexus point for advertisers to insert themselves into. Users who search are searching *for* something. You don't perform a search like "hotels in Marietta Georgia" without having at least some passing interest in booking a hotel in that city in the near future. Advertising around search allowed marketers to reach consumers at the very point of intentionality, at the very moment they were either researching a purchase or actually looking to buy.

An important component of this entire process was the ability to "pay per click" —as opposed to paying based on the number of people who (theoretically) viewed your ad, as every other online advertiser did in the dot-com era. This was the second key innovation: with the GoTo model, an advertiser only "paid for performance." If no one clicked on your ad, you paid nothing. This was a radical but extremely enticing option at a time when click-through rates on banner ads had dropped to minuscule percentages.

Gross had intended for GoTo to become a shopping destination, thus the active tense of the name. And yet, even though advertisers eagerly signed up to hawk their wares, the consumers didn't follow. Undeterred, Gross had the brilliant idea of chasing the traffic he needed. GoTo approached nearly all the extant portals and search engines and offered them what was essentially free money: GoTo would "syndicate" its paid search results so that for almost any keyword on a site like AOL Search, the first three or four results would be GoTo's text links which, though they looked just like the other search results, would actually be ads. When searchers clicked on these paid links, GoTo would share the ad revenue with the portal, thereby instantly monetizing the search traffic.

GoTo succeeded in signing deals with all the major portals, and at a stroke, turned search—which had been a loss leader for portals throughout the nineties—into a cash cow. In 2002, GoTo changed its name to Overture to better reflect its true business model of introducing customers to advertisers. The company was earning more than $78 million dollars a year on $668 million in revenue—all from paid clicks syndicated to the likes of Yahoo, AOL and MSN. Overture saved the portals by fixing a fundamental flaw in their business model. Portals had sprung up in the first place because they needed to be "sticky." None of the early search sites could make money when they sent users out onto the web, so they attempted to hoard the eyeballs and keep them on-site in order to create impressions for banner ads. But now, clicking itself was finally worth something. As the writer John Battelle has put it, Overture could generate billions of dollars, one click, one nickel, at a time.

GoTo/Overture came along at a very opportune moment for the Internet. As the bubble burst and the advertising market cratered, paid search stepped into the breach to replace the lost revenue from all those bankrupt dot-com advertisers. In the case of Yahoo, by the summer of 2002, the paid links it was getting from Overture accounted for more than 10% of the ailing portal's revenue, and almost all of its much-diminished profits.[13] It's not an exaggeration to say that Overture and paid search saved the portals and the search industry in general. Lucky for Google, there was now a very lucrative new advertising model it could copy, and what was more, this new form of ad had proven the immense value of search, Google's crown jewel. But since Larry and Sergey never met an idea they didn't think they could improve upon, Google was not interested in merely copying. If Google was going to have ads, the ads would have to be better than traditional ads; they would have to be *useful*.

■

GOOGLE'S FIRST EXPERIMENT with advertising came in January of 2000 when it began showing unobtrusive text (in keeping with its minimalist aesthetic) links above certain keywords. But the ads were still priced like flashy banner ads, on the traditional CPM (cost per impression) model. Page and Brin wanted something more scientific, more automated. They liked how anyone could buy an ad through Overture by simply using an online form. In October 2000, they launched what they called AdWords, which allowed any advertiser, no matter the size of their operation, to purchase search ads online in a matter of minutes, using a simple credit card.

As GoTo/Overture had discovered, advertisers were eager to get in front of Google's burgeoning search traffic, and the first influx of AdWords advertisers put an end to Google's immediate money issues by bringing in $85 million in 2001. But since the ads remained CPM-based, advertisers were still paying for impressions, not for actual clicks. Google was missing out on the performance-based advertising revolution, and it showed. Overture's 2001 revenues were $288 million, and that number was growing at a faster rate than Google's.[14] In February 2002, Google unveiled a new version of AdWords that copied Overture's cost per click and auction-pricing model. In typical Google fashion, however, its Overture clone had a key innovation that made all the difference in the world.

The new version of AdWords had advertisers bid against competitors' ads, but Google's system was not strictly pay-for-placement. Ever enamored with math and the power of algorithms, Google introduced an important new ranking factor for the ads that it called a "Quality Score." In essence, Google's system took into account how often that ad was actually clicked on, *in addition to* how much an advertiser was willing to pay per click. Each time a search was run and AdWords results were generated alongside the search results, the ranking of the eventual ads decided how relevant the ads actually were. This prevented deep-pocketed but ultimately irrelevant advertisers from dominating every keyword. You could no longer guarantee to rank high just by being willing to pay the most. Your ad also had to be *clicked on the most* in order to rise up the rankings. Successful advertisers paid *less* per click, but ranked higher. If your ad was of good quality, and tended to get clicked on more often, AdWords trusted that it was more relevant for that search phrase and would therefore rank you higher even if you didn't increase your bid. Google did this because, almost counterintuitively, it knew that it stood to make more money when the ads were ranked this way. Over time, more money would come in from a 5-cent ad that was clicked on twenty-five times—than from a dollar ad that was only clicked on once.

From a searcher's perspective, the ads felt less annoying the more relevant they became. To a certain extent, Google's AdWords began to seem almost as useful as the organic search results for certain keywords, because the quality score kept them germane to the searcher's original query. And on the advice of early Google advisor Yossi

Vardi, the bulk of the AdWords appeared on the right-hand third of the search results page. This had the consequence of increasing the amount of ads delivered per search, all while seeming to make them less intrusive. The original, organic search results still filled the main two-thirds of the page, pristine and untarnished. When Google ran limited control experiments where it showed one group of searchers results without the ads, and another group search results *with* the ads, users who saw the ads actually searched more.[15] It became a classic win-win-win: Google started making more money per search than Overture did, advertisers felt like they were paying less per click while reaching more potential customers, and users felt like they were getting supplemental search results, in the form of ads that were often quite useful.

Overnight, Google's fortunes were transformed. Led by a new hire named Sheryl Sandberg (later, more famous for her leading role at Facebook), AdWords became the blockbuster success that Google had been looking for all along. It helped considerably that Google had what Overture didn't: its own highly trafficked search destination. Google didn't have to cut syndication deals with other portals in order to get traffic for its ads, since its own website was already servicing hundreds of millions of searches per day. It didn't *have* to cut deals, but it did anyway, especially a partnership with AOL, announced in May of 2002. Google would not only provide organic search results to AOL, but paid search results as well, stealing the business away from Overture, which had previously provided AOL's paid links. Two thousand two would become Google's first profitable year, with $440 million in sales and $100 million in profits.[16] By 2003, profits were more than $185 million and the AdWords program could boast more than 100,000 advertisers, all without a commensurate rise in Google's head count, because the AdWords sales system was automated.[17]

In retrospect, going into advertising played into Google's deepest strengths. For a company full of data-obsessed nerds, Google looked at advertising as just another problem smart algorithms could solve. Indeed, serving the appropriate ads alongside the organic results, running auctions in real time for billions of searchers, and reranking the ads according to their performance became an even more complicated algorithmic trick than even search had been. But then, Google's entire infrastructure was devoted to crunching numbers and organizing vast amounts of data, so it was uniquely positioned to get this sort of thing exactly right. Just as with web search, when Google turned on its new advertising algorithms, it found that the ads got better over time; so much so that Google's computers could eventually predict with stone-cold accuracy which ads would work and which wouldn't.

Google can be thought of as a company born from two miracle inventions, one of which it came up with itself, and the other of which was cribbed from Overture. Definitively solving the problem of web search is obviously the miracle that has made the largest impact on our society. The web and the Internet itself are now so big that without decent search, it's easy to imagine that the whole edifice would have collapsed under its own complexity by now. But by improving on Overture's pioneering work with paid links, Google was able to achieve something just as amazing: it made the Internet profitable at scale and for the first time. Paid search would prove to be the greatest advertising engine yet devised by man. Furthermore, algorithmically served ads would support nearly every product Google would release subsequently: Image Search, Google News, Gmail, Google Maps, Google Books. In a few short years, search ads would surpass traditional banner or "display" ads, and within a decade, Google would be generating more than $50 billion in revenue,[18] having captured nearly 50 cents of every dollar spent advertising online. Today, most advertising is automated in ways similar to what Google pioneered, and even now the largest market for online advertising remains tied to search. It turned out that the gold mine on the Internet was search all along, as Yahoo and others had first intuited, but had subsequently forgotten.

■

BY 2003, GOOGLE WAS OBSESSED with one thing: keeping all this a secret. As ever, Google feared tipping Microsoft off to the value inherent in search. Sure, Microsoft was ailing from the antitrust trial and was already entering its lost decade, but it was still the only technology company that had the resources, talent and size to do to Google what Google had done to Overture.

Helping to keep Bill Gates and company in the dark was Google's new "grown up" CEO, Eric Schmidt. Schmidt had been a longtime Microsoft adversary going back to the 1980s, when he was an early manager at Sun Microsystems, and then briefly in the 1990s as CEO of Novell. Years of experience managing a relationship with Microsoft no doubt played a role in Schmidt's eventual selection as CEO, but a willingness to swallow his ego was probably what put Schmidt's candidacy over the top. Becoming the Google CEO meant having to share the limelight—as well as some degree of the decision-making process—with Google's founders. Indeed, the working relationship Schmidt would go on to form with Page and Brin evolved into a sort of triumvirate where all three had meaningful say. Though, if push came to shove, the founders could outvote the CEO. Page and Brin's dream candidate for the job had been Steve Jobs, but it's hard to imagine the Apple founder being willing to take a back seat to two twenty-seven-year-olds, as Schmidt eventually agreed to do.

Capable management was crucial as competitors circled. Thanks to its investment in Google, Yahoo had the best inkling as to what was really going on behind the scenes at the Googleplex. In the summer of 2002, only a few months after the new version of AdWords debuted, Yahoo made a $3 billion bid to buy Google outright. Google, with Schmidt newly at the helm, turned down the offer. Too late, Yahoo realized that search was the motherlode of business models, so it canceled its organic search partnership with Google, purchased what was widely considered to be the company with the second-best search technology, Inktomi, for $257 million, and in 2003, paid $1.4 billion dollars to acquire Overture. The idea was to combine the two properties under the Yahoo umbrella and replicate Google's algorithms-and-advertising juggernaut, complete with a quality score and bidding systems that mimicked AdWords in efficiency and effectiveness. Called Project Panama, this next-generation system was not released widely until February of 2007, by which point Google had run away not just with search market share generally, but virtually the entire search advertising market.

By then, the whole world knew what Yahoo had intuited: Google was printing money. On April 29, 2004, Google filed for an initial public offering of stock. It would be the highest-profile technology IPO since the dot-com bubble burst. When Google released a snapshot of its financials so that potential investors could evaluate the company's prospects, both the technology and financial worlds were amazed. Venture capitalist Mitchell Kertzman told the *Wall Street Journal* that Google's numbers were "stunning."[19] Google's PR head David Krane remembered the general response being " 'Holy shit!' "[20] Google had generated more than half a billion dollars in cash flow in 2003 and its operating margins stood at an astounding 60%. These were Microsoft-level numbers.[21] The online market for search ads had reached $2.5 billion in 2003 (nearly tripling the size of the market from the $927 million spent a year before), and Google had captured approximately $1 billion of that.[22] A lot of this success was thanks to the fact that 35% of all web searches were now being done through Google, surpassing Yahoo's 30% market share for the first time.[23]

Brin and Page had not actually wanted Google to go public, having filed only because financial rules put into place after the dot-com bubble burst would soon compel them to do so. In the letter the founders wrote to prospective investors, which they called " 'An Owner's Manual' for Google's Shareholders" (and which the *New York Times* declared to be "part financial document, part populist manifesto")[24] Google's founders began with a simple statement: "Google is not a conventional company. We do not intend to become one."[25] Brin and Page went on to state their intention to continue to operate Google in the service of their own lofty ideals, to "develop services that improve the lives of as many people as possible—to do things that matter" rather than bow to the quarterly whims of Wall Street's expectations. Throughout the coming months, as the ramp-up to the IPO began, the Google guys were accused of "thumbing their nose" at Wall Street and its traditions.[26] Larry and Sergey demanded that the underwriters of the IPO receive a fee of only 2.8% for their services, about half the rate bankers usually expect.[27] During the "road show" when the founders crisscrossed the country, ostensibly to sell the company to investors, Larry and Sergey drew fire for flat-out refusing to answer specific questions about Google's operations or future plans.[28] Even the amount of shares Google was offering to the public was a bit of a prank. Google wanted to sell exactly $2,718,281,828 worth of equity. Math geeks (like the Google founders) knew that this number represented the first 9 decimal places in the mathematical number *e*, which is, of course, an irrational number.[29]

On August 19, 2004, Google went public at $85 a share, and rose 18% on its first day of trading, to close at $100.34. The 38 million shares that Larry and Sergey each held in the company were worth approximately $3.8 billion at the close.[30] Google was valued at $27 billion,[31] more than a bit behind Yahoo's $38.7 billion market cap. But that disparity wouldn't last long. By the time Google's first quarterly report as a public company revealed that sales had doubled from the previous year, Google stock passed $200.[32]

It is impossible to overstate how important Google's IPO was to the Internet, Silicon Valley and the stock market overall. As the *New York Times* said on the day after the company filed to go public, it was "as if the dot-com glory days never ended."[33] Google's success was validation that the Internet as a social, cultural, and (most important) a financial phenomenon was not dead. The revolution had merely been regrouping. Google was also proof that not only were some of the original ideas from the dot-com era still valid; some new ideas might also be out there ready to build on the dot-com era's faded promise. Within Google itself, there were whispers of exciting new projects, like, some sort of a Google "phone" so that searchers could get answers to queries at any moment no matter where they were.[34] More than anything, Google's success provided the template to make these new ideas profitable. Just as with the Netscape IPO nearly a full decade before, a new generation took notice: there was fire in Silicon Valley again.

■

GOOGLE'S UNORTHODOX TRANSFORMATION into an advertising juggernaut had further, unexpected outcomes. Millions of small and medium-sized businesses eagerly signed up, advertisers who, in a previous era, might have paid for an ad in the Yellow Pages or taken out space in a local newspaper's classified section. Now they were able to design and implement advertising strategies that had the same global reach as the web itself. The erosion of traditional advertising channels that had begun because of sites like eBay began to accelerate in the first half of the 2000s.

This digital economy didn't just flower on the marketing side of the equation, because Google had developed a way to monetize content as well. This was AdSense, which Google launched soon after AdWords. Google engineers dreamed up ways to syndicate text ads not just to major search sites and portals, but to the entire web itself. "The idea of putting ads on nonsearch pages had been floating around here for a long time," Google executive Susan Wojcicki said later. Google already had basically the entire web in its index, so if it could find a way to match relevant ads to the content on other people's web pages (just as it had matched relevant ads to search queries), Google could, in Wojcicki's words, "change the economics of the web. You do the content and leave the selling of the ads to Google."[35]

When Google announced in February of 2003 that it was purchasing a small company named Pyra Labs, a lot of people were confused. In August of 1999, Pyra released a software program to help people "blog"—a phenomenon that was becoming popular at the tail end of the nineties. But then the bubble burst, venture capital dried up, and despite the fact that in one year, Pyra's blogging platforms (Blogger, and later, Blogspot) went from hosting 2,300 blogs to 100,000 (and 700,000 a year after that), the company was on life support.[36] Pyra's cofounder Evan Williams (who would go on to be a cofounder of Twitter) laid off every employee of the company except himself and continued to run the sites on his own computer, on his own dime, from his own home.

Google swooping in and rescuing Blogger seemed odd. Pyra Labs was a failed (failing?) company. Blogging was a new phenomenon that smelled very much like a fad to a lot of people. Pundits speculated that Google simply wanted Blogger to improve its algorithms. When AdSense was announced soon afterward, it suddenly all made sense: Google was now in the business of monetizing content on the web, and the long tail of content generated by the millions of blogs coming online from sites like Blogger and Blogspot would be the quickest way to scale up rapidly. It turned out that blogging represented the vanguard of a new kind of web, one that built off the original promise of the web as an interactive medium, but now in a new, more personal way. There was a whole new world of content being created on the web, and the creators were the web users themselves.

■

THE ROOTS OF BLOGGING are obscure. Perhaps the earliest version of the format came from a programmer at the University of Pennsylvania, Ranjit Bhatnagar, who, beginning in November of 1993, started posting, in reverse chronological order, what he had for lunch every single day at "Ranjit's HTTP Playground." Credit for coining the term "weblog" is generally given to the site Robot Wisdom WebLog, launched on December 17, 1997, by Jorn Barger. Shortening the term to "blog" is often attributed to Peter Merholz, who ran a personal website at Peterme.com.[37] But it's not entirely clear when simply publishing a webpage or a "homepage" morphed into publishing a "blog." Ever since the web was born, the idea of webpages as individual soapboxes was one of the most obvious and enticing use cases for the technology. It all tied into the original utopian ideal of the web: anyone with an opinion or an insight could broadcast their truth to the entire planet, free from the oversight of the traditional gatekeepers who told you what you could and could not say. But blogging was somehow more personal and more purposeful than simply having a homepage. The whole point of having a blog was to share something with the world, anything from links to things you found cool, to the most intimate details of your life, to your manifesto for world peace. As Merholz himself said of the blogs, "These sites (mine included!) tend to be a kind of information upchucking."[38]

Justin Hall was one of the earliest "proto-bloggers." On January 22, 1994, when he was just nineteen years old, Hall set up his own personal webpage, eventually named Justin's Links from the Underground, using his student Internet account at Swarthmore College. More than most early web publishers, Hall's subject matter for his website was himself: links, diary entries about his love life, gossip, pictures of his genitalia, etc. In 1994, he begged his way into an internship at *HotWired* and was present for the launch of that pioneering website. While at *Wired/HotWired*, he fell in with the Suck.com crew, who encouraged him to post to his website daily, as Suck was just then attempting. Hall took up the challenge and for an entire decade, nearly daily, links, photos, musings, diary entries, correspondence with readers, personal triumphs and mental breakdowns, all flowed freely on Hall's homepage. Like Ranjit Bhatnagar, Hall felt there was nothing too personal or too mundane to share—even lunch. "It's so much fun," Hall would write, "putting everything out there."[39]

Dave Winer was a veteran software developer who simultaneously became entranced by the web's promise of unfiltered honesty and discourse. Proprietor of a popular technology email discussion list read by industry insiders including Bill Gates, in October of 1994, he moved his musings to the web at DaveNet (eventually, Scripting.com). Like Justin Hall, Dave Winer was in love with the democratizing platform the web provided. "Imagine being able to find out what's (*)really(*) going on in anyone's life. What if everyone wrote about their issues. We could all learn from each other."[40] Winer became a vocal proselytizer for using your personal platform to engage; not just to publish, but to share, debate, argue, respond, provoke and question. DaveNet was his own personal soapbox, but he encouraged others to launch their own soapboxes. And because he was a gifted programmer, he was able to do more than just encourage others, he gave them the tools to do so as well, creating software programs like the NewsPage Suite, Radio UserLand, and Manila. These programs helped people set up their own soapbox-like websites, and helped formalize conventions we now understand as "blogging" like the reverse chronological format of updates, webrings and blogrolls to link to likeminded sites and the ability for readers to post comments on posts. Most crucially, he helped advance and popularize RSS (short for Really Simple Syndication), which helped bloggers alert the world when they had published something new.

It was the new publishing tools, like Winer's and like Blogger's (eventually, there would be many more, such as Moveable Type, LiveJournal and WordPress), that really helped the medium of blogging take off. Even though creating a website was relatively simple from the very first days, publishing on the web still required some level of technical nous. Thanks to the explosion of blogging software, by the late nineties, you could push a button and, presto, you were published online.

Matt Drudge was a twenty-eight-year-old sales clerk in the CBS studios gift shop in Hollywood when, in 1994, he launched an email newsletter focused on Hollywood gossip, some of which he overheard on the CBS lot, and some of which he later admitted had been pilfered from CBS's mailroom wastebaskets. The newsletter evolved into a blog, because Drudge intuited that the web provided him with a platform that was as powerful as any news organization in the world. "I have no editor," he would later tell *Newsweek*. "I can say whatever I want."[41] In a speech to the National Press Club after he had become world-famous, Drudge declared: "With a modem, anyone can follow the world and report on the world—no middle man, no big brother."[42]

Drudge gained his notoriety in January 1998 when, after *Newsweek* had determined the story too dubious to publish itself, he released the first rumors about Bill Clinton's affair with a White House intern on the Drudge Report. One man's digital soapbox nearly brought down the President of the United States. Within 6 months, the Drudge Report claimed 6 million monthly visitors, which represented a greater readership than *Time* magazine.[43] By 2007, with the help of a single employee by the name of Andrew Breitbart (later, founder of Breitbart News), DrudgeReport.com made millions of dollars a year from advertisements on the site from the likes of the *New York Times*.[44]

Matt Drudge's ascendance into the top ranks of newsmakers and publishers caught the attention of other Internet-savvy hustlers. Nick Denton had been a journalist at the *Financial Times* in London when the budding blogging scene caught his fancy. He began posting nearly daily on his own NickDenton.org. "You could express yourself," he said of blogging's simple appeal. "I could express opinions."[45] Feeding off the newness of blogging, and referencing the Fleet Street–style tabloids from his native Britain, as well as satirical publications like *Private Eye* and *Spy*, Denton launched a series of blogs under the umbrella company named after the first one to debut: Gawker.

Launched in 2002, Gawker was a straight-up tabloid, covering the foibles of the New York media industry. "Nick had the brilliant insight that if you want to get people to read something, the easiest way is to write about them," remembered Lockhart Steele, another early blogger whom Denton would eventually hire into the Gawker stable of writers.[46] But, it was Gawker's voice and attitude, its much-commented-upon "snarkiness" that really drew attention. Gawker had a habit of commenting on the news broken by other publications, linking to published pieces and offering commentary on them. Gawker also critiqued other publications themselves, often with viciously biting commentary. The editorial attitude of Denton's publications drew a lot from the spitballing-from-the-back-of-the-classroom Suck.com. Indeed, when Denton launched the blog Wonkette, to lampoon the Washington, D.C., establishment, he hired a Suck alumna named Ana Marie Cox to do so.

"EXCLUSIVE: The Condé Nast cafeteria"

Filed to: Condé Nast

Gawker had reported previously that the Hamburger Guy in the cafeteria had been fired after impatiently tapping the glass partition between himself and The Anna [Wintour, legendary editor in chief of *Vogue*] *an act of insolence not to be repeated by any cafeteria slave wishing to end his or her day in the employ of Si Newhouse, Jr. Not so, said the mole. "He just wanted to learn how to make pasta, so they moved him."*

GAWKER, MARCH 24, 2003, 1:22PM[47]

Soon, Denton had his portfolio of blog publications covering a universe of topics from personal productivity (Lifehacker) to Silicon Valley (Valleywag) to video games (Kotaku) to sports (Deadspin). Denton kept expenses low, paying his bloggers a couple thousand dollars a month (at best) but expecting a dozen or more posts from each blogger, each day. By providing always new, always updating, always up-to-the-minute content, people would return again and again to Gawker's blog feeds to find out what was going on in the world. "Immediacy is more important than accuracy," Denton would say, "and humor is more important than accuracy."[48] Traditional journalists would scoff at the shoddy editorial standards of blogs like Denton's, but they couldn't argue with the way blogging began to drive the daily conversation in ways that traditional publishers couldn't. By keeping his expenses low and taking advantage of the new advertising technologies like AdSense, Denton created a media empire one blog at a time. By 2007, Gawker had grown to around 100 employees and $10 million to $12 million a year in annual profits.[49]

14

WEB 2.0

Wikipedia, YouTube and the Wisdom of Crowds

In a way, blogging was simply the inevitable migration of publishing into the digital arena. The music site Pitchfork.com, which flourished in the early 2000s, was simply doing things that magazines like *Spin* and *Rolling Stone* had been doing in paper form for years: reviewing music and profiling new artists. But *Pitchfork* also encapsulates how blogging changed the media landscape in terms of taste-making and authority. *Pitchfork* allowed a slate of obscure music writers to challenge the established order merely by gaining credibility through the power of their unique point of view. This phenomenon, whereby the best content rose to the top and the most prominent voices became the new "establishment," occurred in numerous interest niches across the Internet. From food to fashion, from automobile blogs to "mommy" blogs, even touching such rarefied academic arenas as finance, economics and the law, blogs allowed new voices to surface and claim the mantle of "expert," without any official sanction, training or even previous experience.

Perhaps the most illustrative example of this came in the realm of politics. September 11, 2001, was transformative for obvious reasons. But that tragedy was also the first time a historical event could be recorded online from the perspective of those who experienced it firsthand. Thousands of bloggers recorded their emotions and their impressions and even their direct experiences for posterity. "Only through the human stories of escape or loss have I really felt the disaster," Nick Denton wrote for the *Guardian* newspaper on September 20, 2001. "And some of the best eyewitness accounts and personal diaries of the aftermath have been published on weblogs."[1] It was what Justin Hall had been advocating for years: the common man as recording vessel for history. "If everyone was to tell their stories on the web, we would have an endless human storybook, with alternating perspectives. . . . Give someone a digital camera, a laptop, and a cellular telephone, and you've got an on-the-spot multimedia storyteller from anywhere in the world."[2]

From the right side of the U.S. political spectrum, the response to 9/11 was immediate and strident. A group of conservative-leaning blogs like Instapundit, Little Green Footballs, Power Line and others, began advocating for an aggressive global war on terrorism. These sites were known collectively as the "war bloggers" in the coming years as they became vociferous cheerleaders for the wars in Afghanistan and Iraq. Conversely, it was opposition to the Iraq War that saw a community of left-wing blogs spring up like MyDD, DailyKos, Eschaton, Hullabaloo and more. The lefty blogosphere called itself the "netroots" and could rightfully claim credit for giving energy to the brief, insurgent antiwar presidential candidacy of Howard Dean in 2003.

Again, on both the left and the right side of the "blogosphere," new voices rose from seemingly nowhere, gaining a reputation through smart comments posted on popular blogs, graduating to influential blogs of their own, and then often going on to positions of prominence at "mainstream" journalistic publications or even actual political positions. In the United States, we live in a post–political-blogging world where movements can arise online and take over the mainstream discourse. The most prominent examples of this new reality come from the right, in the form of the Tea Party movement and especially the Trump presidency, which has seen bloggers (in the form of Breitbart) ascend to the highest corridors of political power.

But perhaps what's most interesting to observe about the rise of blogging is how the habits and behavior of web users themselves changed. If the web in the dot-com era had been about, in the words of the technology journalist Sarah Lacey, "taking prepackaged content from the offline world and throwing it onto a site,"[3] the new web was about you (and everybody else) putting up your own content, discovering it for yourself (and others), organizing it yourself and determining that your content was just as interesting and valuable as anything else in the media landscape. It had taken about a decade for mainstream users to acclimatize themselves to the web, but now that they had the lay of the land, they were no longer content to merely "surf." Even everyday web users were now ready to *participate* in the web. As Marc Andreessen had anticipated all the way back in the days of the Mosaic browser, the "riff-raff" were ready to join the party in a major way, not just as consumers, but as producers. To quote the title of a popular book from shortly after this era, the postbubble Internet was a moment of *Here Comes Everybody*.

Some credit can be given to Napster for opening these floodgates. All those tens of millions of users who traded MP3 files were proactively and spontaneously self-organizing and using their own libraries to create content for others. Napster was the first time mainstream web users saw the utility in producing, not just consuming, content. And baked into Napster was a "social" component to all this activity. If you found a song you liked from another user on Napster, you could also browse the other files in that user's library. If you both shared an interest in a given band, then maybe you would like that *other* band that your friend on Napster had so many MP3s of. It was like the Netflix recommendation engine, but impromptu and self-created. It was the act of finding like-minded individuals, of creating community out of silos of shared interest.

This "social" aspect of the web began manifesting itself in a number of ways beyond Napster and blogging. A link-blog site called Slashdot grew popular around the turn of the century by aggregating the blog-post and news-item deluge that came online every single day. In the comments of every link post, the thousands of members of the Slashdot community debated and discussed the posted articles. Order was given to the chaos by the Slashdot community itself. Randomly selected users were given moderation privileges to vote up or vote down content on a scale ranging from "insightful" to "troll," thereby allowing the community to police discourse on its own.

Digital cameras were just becoming popular in the early 2000s, and sure, you could make actual prints from your photos on your ink-jet printer and then mail those to your grandmother; but conversely, you could also just post an entire album online via a site like Flickr (launched in February of 2004) and simply send Nana the link to your Flickr page. More than that, you could share your pictures with complete strangers if you wanted to. How would strangers find your photos? Well, Flickr allowed you to "tag" your photos with keywords that enabled other users to search for them. If someone wanted to browse a bunch of photos of the Grand Canyon, they could type those keywords into Flickr and see the results of a thousand different strangers' summer vacations.

From the days of the Netscape browser, users had used bookmarks and "favorites" to keep track of their favorite web pages. But what if you wanted to see what other people had bookmarked? Del.icio.us (launched in September of 2003) let you do just that, allowing users to discover cool new things on the web by sharing their bookmarks with each other, just as Napster had allowed them to exchange songs.

The new postbubble web was about the users and the content in equal measure. It was about spontaneous impulses like "sharing" and self-organizing schemes like "tagging" and taxonomies. It was about how the content created by and for the hoi polloi often ended up being more engaging and exciting than the content that was prepackaged or professionally produced. And increasingly, the new web was about the collective "wisdom" of the crowd to create and organize the anarchy.

The idea of collaborative effort and collective organization had long been a common practice in hacker and software development circles. Just as each of the hackers on woowoo had pitched in to help Shawn Fanning refine Napster, groups of programmers often came together and formed communities around the development of "open source" projects like the Linux operating system. Far from being a case of "too many cooks in the kitchen" creating a muddled fiasco, open-source development proved that complete strangers could independently, and without much centralized coordination, come together to collectively produce things in an orderly, sublime way.

A veteran software developer named Ward Cunningham brought this practice to the web for the first time on his Portland Pattern Repository, a website for other programmers to contribute and share programming ideas. On March 25, 1995, Cunningham installed a subpage on the site called WikiWikiWeb. The "wiki" (the term came from the Hawaiian word for "quick") constituted a series of pages that could be edited by any user. So, a given user might post some code patterns to the wiki, and another user might come behind him and add to those patterns, change them, even completely replace them. But all edits were stored, and the page could revert to previous versions if any user chose to do so. It seems counterintuitive that such a system could work, but Cunningham learned that, given enough input from enough interested users, his Wiki system worked quite well. Cunningham is famous for coining "Cunningham's Law," which finds that "the best way to get the right answer on the Internet is not to ask a question, it's to post the wrong answer."[4] If a user contributed code patterns to his site that other users found wrong or merely objectionable, Cunningham found that, almost inevitably, another user would come along and right the wrong.

Wikis tapped into a powerful impulse of collective action. A few years later, an obscure entrepreneur would make use of this impulse to save his own struggling creation. Jimmy Wales was a serial dot-com entrepreneur who had found a modest degree of success by creating more sophisticated web directories—sites like Yahoo, but more focused. Wales also had a lifelong passion for encyclopedias and was obsessed with the notion that the web could create the largest encyclopedia conceivable. "Imagine a world in which every single person is given free access to the sum of all human knowledge," Wales would write later.[5] In early 2000, he launched what he called his Nupedia project, soliciting experts in a wide range of fields to contribute articles for what he hoped would eventually become an infinity encyclopedia. Contributors to the project were required to be knowledgeable in a given topic, and they would have to submit their articles to a rigid system of peer review by vetted editors. Also, the editors themselves had to be credentialed. "We wish editors to be true experts in their fields and (with few exceptions) possess Ph.D.s.," the Nupedia policy stated.

But Nupedia's rigid quality control apparatus proved inefficient. It wasn't until September 2000 that the first article made it through the layers of editors, and by the end of the year, less than two dozen had been published on Nupedia's website. In frustration, on January 10, 2001, Wales installed a descendant of Cunningham's original wiki software on Nupedia's server. This "Wikipedia" was merely intended as a separate feeder service to speed up the Nupedia submissions process. Articles would be collectively written and edited on Wikipedia, then fed into the existing peer-review editing process. Almost immediately, however, Wikipedia overtook Nupedia not just in the quantity of articles that were created, but in the quality as well. The first article created, on January 15, was on the letter "U" and investigated the origins and usage of the twenty-first letter of the English alphabet.[6] It was comprehensive, it was well written, and it was—to the surprise of Wales and his team of editors—accurate. The few thousand users who had shown up to test out Wikipedia had, through their collective input and edits, gotten the article polished to near-authoritative quality.

Within a month, Wikipedia had around 600 articles, achieving in a matter of weeks more than Nupedia had achieved in a year. The experiment was promoted on Slashdot, and soon Wikipedia was flooded with Slashdot's passionate users, members of a community who were already comfortable with collective editorial action. Within a year, Wikipedia had grown to 20,000 articles. By 2003, the English-language Wikipedia had more than 100,000 articles, and versions of the service were springing up in every language imaginable. By that point, Nupedia and its rigorous system of editors and peer review had long been abandoned.

What confounded everyone who learned of the success of Wikipedia was that it actually worked! "Couldn't total idiots put up blatantly false or biased descriptions of things, to advance their ideological agendas?" asked one of the leads of the original Nupedia project on internal Wikipedia message forums. "Yes," replied a Wikipedia partisan, "and other idiots could delete those changes or edit them into something better."[7] It turned out that the "infinite monkey theorem" about giving enough monkeys typewriters and eventually producing Shakespeare—was not exactly fanciful. Enough self-interested strangers could achieve a fair degree of accuracy on a wide range of topics. In 2006, there were 45,000 active editors of the English-language version of Wikipedia alone.[8]

And Wikipedia had unique advantages that the web made possible. In the coming years, as any news or historical event occurred, Wikipedia contributors would post an up-to-the-minute factual summation of these events, and then amend the entries to reflect changing circumstances or new information. Wikipedia was often accurate and authoritative in near-real time, and it had the infinite space and resources of the Internet to play with, so it could serve what became known as the "long-tail" of content. Any encyclopedia worth its salt might have an article on World War II. But Wikipedia could produce a 418-word entry on, say, the Compton railway station, an abandoned stop on the Didcot, Newbury & Southampton Railway in England. Or, it could produce a detailed plot and development synopsis on Season 8, episode 14 of the TV show *Cheers*, the one where Cliff Clavin goes on *Jeopardy*. No other encyclopedia in history was capable of that sort of breadth of topics.

Wikipedia was a modern miracle and soon became one of the most trafficked websites in the world. Wales had originally intended the project to be a commercial one, supported by advertising. But when the contributors and editors revolted at the very suggestion of putting ads up on Wikipedia, Wales instead made the site into a nonprofit enterprise. To this day, it is supported by contributions from the public and is thereby an open-source counterweight to the proprietary "answer engine" that is Google.

■

GRADUALLY, PEOPLE BEGAN to notice that there was a new energy on the web and it shared several characteristics. The long tail. The wisdom of crowds. Users creating content of and to their own design. In 2004, this new Internet energy gained the name Web 2.0, after a similarly named conference held in October 2004. If Web 1.0 was about browsing stuff created by others, Web 2.0 was about creating stuff yourself. If Web 1.0 was about connecting all the computers in the world together, then Web 2.0 was about connecting all the *people* in the world together, via those interlaced computers. If the clarion call of Web 1.0 was the Netscape IPO, then the coming of age of Web 2.0 was Google's IPO. "Web 2.0 means using the web the way it's meant to be used," wrote Paul Graham, a veteran entrepreneur of the Web 1.0 era who would soon become a key driver of Web 2.0 as an investor. "The 'trends' we're seeing now are simply the inherent nature of the web emerging from under the broken models that got imposed on it during the Bubble."[9]

Within the technology industry itself, the sense that the Internet revolution was back in gear came via the promotional efforts of—what else?—a blog. On June 10, 2005, Michael Arrington, a thirty-five-year-old former Silicon Valley lawyer who was active during the dot-com years, started posting to a personal blog at TechCrunch.com. Arrington's entries were mostly musings about the new services, websites and companies he saw bubbling up through the Web 2.0 scene. But he soon branched out to covering the actual news of Web 2.0: what new companies were being founded and by whom; what startups were raising an investment round and with whom; what hot new websites had been acquired, and by whom. TechCrunch became not only the cheerleader of the Web 2.0 movement, but, in a sense, proof that the movement even existed. Arrington simultaneously became a power player in his own right, since his site became a PR bonanza for whatever new service or company he deigned to blog about. As *Wired* magazine put it, "A positive 400-word write-up on TechCrunch usually means a sudden bump in traffic and a major uptick in credibility among potential investors." When TechCrunch gave a glowing write-up to a startup named Scribd, as *Wired* reported, "CEO and cofounder Trip Adler says he had 10 calls from venture capitalists within 48 hours."[10]

Indeed, the startup scene was back in full swing, in no small part thanks to TechCrunch and the hype around Web 2.0. Usage of the Internet had never dipped and indeed was finally reaching critical mass in the developed world. In 2003 alone, the percentage of Americans with broadband Internet connections in their home increased from 15 to 25%.[11] A new technology called WiFi arrived on the stage to make the notion of surfing the web something that felt ubiquitous and commodified. Even online advertising was coming back, providing that same old business model (but with different tools and greater numbers) to new online efforts. Between 2002 and 2006, U.S. advertisers increased their online ad spend from $6 billion to $16.9 billion.[12]

The venture capital machine started to lurch back into life to fund this new activity. VC investments in U.S. startups bottomed out at $19.7 billion in 2003, a far cry from the dot-com–era peak of more than $100 billion in the year 2000.[13] In the coming years, VC investment would rise—modestly but steadily—reaching $29.4 billion in 2007.[14] A slew of new companies were funded, but the renewed interest in Internet startups was not a replay of the late-nineties frenzy. Both investors and entrepreneurs had been chastened by the bubble's aftermath. Get Big Fast was no longer the strategic mantra; multimillion-dollar advertising campaigns and gaudy launch parties were out. Instead, Web 2.0 companies aimed at refining their products and services, carefully cultivating a user base through feature innovation and word-of-mouth discovery, all while focusing like a laser on issues such as reliability and scalability.

VC investment didn't roar back in huge numbers because it didn't have to. In the Web 2.0 era, you could create a service used by millions in a matter of months, and you could do so for pennies on the dollar—at least, compared to the dot-com era. The hangover from the bubble fallout meant that talented programmers could be hired on the cheap; the infrastructure glut leftover from the global fiber buildout meant that bandwidth, storage and data costs were lower; and the tools developed during the bubble meant that you didn't have to build a company from scratch anymore—you could cobble one together using free and open-source tools to assemble the building blocks of a minimum-viable product for next to nothing. By some estimates, the cost of starting a web company had fallen by 90% in the few short years of the nuclear winter.[15]

The website Digg was perhaps the poster-child company of the Web 2.0 era, and it illustrates this change in startup economics perfectly. In 2004, twenty-seven-year-old Kevin Rose had an idea for a new website that would help plugged-in geeks like himself discover the news of the day: whatever was hot on the blogs or even mainstream sites like the *New York Times*. His vision was of a site that took the community-voting aspects of Slashdot, but gave the power to surface news to anybody. On Digg.com, any user could submit a story and other users could "digg" it. If enough users dugg, then the story would rise to the front page. Conversely, if users didn't like a story, they could vote to "bury" it. Rose registered the Digg.com domain name (that was the biggest expense, actually; he had to buy the domain from an existing owner), paid a programmer in Canada $12 an hour to code up the site, and paid $90 a month to have a company host it. The site launched on December 5, 2004.[16] Rose's total outlay was around $10,000.

For that investment, Rose soon had the hottest site on the Internet. Within a year, Digg passed Slashdot in traffic.[17] Making it to the front page of Digg could drive scads of traffic to a website, so publishers all around the web began to add "Digg This" buttons to their websites. Within two years, Digg had nearly as much web traffic as the *New York Times* and more than 1 million people came to the site daily, "digging" thousands of stories.[18] Digg was nominally profitable from day one, thanks to AdSense ads from Google, and later, banner ads from more traditional marketing networks. In 2007, Digg landed a $100 million ad deal with Microsoft. By that point, Rose had appeared on the cover of *Businessweek* under the headline "How This Kid Made $60 Million in 18 Months." That estimation of Rose's paper wealth came from the valuation given to Digg by venture capitalists. But the truth was, Digg had only raised money reluctantly. As Rose and his cofounder Jay Adelson made the rounds on Sand Hill Road, home to the most powerful Silicon Valley VCs, they were shocked by what they saw as the outdated thinking among the money men. "They are still back in the 1998 belief system that it's all about the portals," Adelson marveled.[19] The VCs wanted to throw tens of millions of dollars at them in order to build the next Yahoo or AOL. Rose and Adelson were content to raise a paltry $2 million. They didn't really need the funding, and besides, raising less money meant keeping more equity for themselves.

The new Web 2.0 companies didn't need to raise as much money and, unlike just a few years previously, none of them were in any hurry to go public. In the wake of the bubble bursting, a wave of scandals involving companies such as Enron and WorldCom had ushered in a new era of financial regulations. The Sarbanes-Oxley legislation especially meant that there were fewer advantages to going public and more incentives to stay private for as long as possible. Without the venture capitalists breathing down their necks for a financial "exit," the Web 2.0 companies were more in control of their own destinies and wary of the pressures that a blockbuster IPO would impose upon them. The lesson of the bubble had been learned: you can go for broke, but try to build a *real* company first.

That didn't mean the money men were denied their "exits." As the survivors of the dot-com bubble began to see their balance sheets return to health, there was an entire group of deep-pocketed acquirers that would begin to pick off the most promising members of the Web 2.0 class. Yahoo swallowed up Flickr and Del.icio.us in 2005, for around $40 million and $20 million, respectively. Scandinavians Niklas Zennström and Janus Friis created the second-generation peer-to-peer networking platform Kazaa before turning to that same P2P technology in order to make phone calls over the web. They founded Skype, enabled hundreds of millions of users worldwide to call and chat with each other for free, and sold the company to eBay for $2.6 billion in September 2005.

But the acquisition saga everyone followed in those early Web 2.0 days was that of YouTube. Late in 2004, three former PayPal employees, Chad Hurley, Steve Chen and Jawed Karim, were mulling over a problem: why wasn't it as easy to post a video to the web as it was to post a photo to Flickr or a blog post to a blog? YouTube was the site they launched to solve that problem, and from the very beginning, the overriding idea was for dead-simple, push-button video uploading.

But what, exactly, should people be encouraged to upload? Should YouTube encourage people to create original, dramatic videos with near–television-production quality? Or maybe YouTube would just host videos for eBay auctions and use the thriving auction economy to jumpstart growth just as PayPal had (they were card-carrying members of the PayPal Mafia, remember). There was even some early discussion about copying HotorNot.com, a popular Web2.0 site where users uploaded profile pictures, and other users voted the portraits up or down based on attractiveness. "In the end, we just sat back," said Hurley, meaning they just let the users upload whatever they wanted no matter how silly, or inane, or personal, or *whatever.*[20] It was the Web 2.0 way.

The first video posted to YouTube exemplified this attitude. *Me at the Zoo* is a nineteen-second video of Jawed Karim at the San Diego Zoo in front of the elephant exhibit. Uploaded on April 23, 2005, Karim offered the following pithy narration:

> Alright, so here we are in front of the, uh, elephants. Uh. The cool thing about these guys is that they have really, really, really long, um, trunks, and that's, that's cool. And that's pretty much all there is to say.

Not exactly "one small step for man" stuff, but credit to the YouTube guys for understanding that that was exactly the sort of video that YouTube was good for.

YouTube was fortunate in its timing. By 2005, broadband Internet adoption continued to increase, and consumer video cameras were becoming common. Even some cell phones allowed you to shoot video by the time YouTube launched. In August of 2005, YouTube got favorable coverage from TechCrunch as well as Slashdot. The number of videos posted started to increase. And then, the post-anything spirit of blogging that YouTube was mimicking helped traffic ramp up even more. In fact, it was the blogs themselves that really helped YouTube explode in popularity. The blogs—and social networks like Myspace.

Aside from push-button-easy uploading, the true brilliance of YouTube was the site's second important focus: dead-simple sharing. After you posted a video to YouTube, you could simply share a link to your uploaded video, just like with Flickr. But you could just as easily cut and paste a few lines of code and your video would play, embedded, wherever you wanted it to: on your website, your blog, or your Myspace page. You didn't ever have to send people to YouTube if you didn't want to. Suddenly, videos were popping up all around the web at a time when web video was still a relatively rare phenomenon. Every time someone embedded a video on a random website, there was that little YouTube logo at the bottom that encouraged people to visit YouTube and try posting videos themselves.

YouTube was incredibly popular on Myspace, but it was the combination of Myspace and the blogs that really caused YouTube to take off. It was the "share-yourself, share-anything!" ethos of the moment combined with the ubiquitous distribution platform of the web that led to what we now call "virality." This was proven by the smash online success of the *Lazy Sunday* video. In 2005, *Saturday Night Live* aired a roughly two-minute musical skit chronicling the antics of a couple of young white dudes in Manhattan hitting up Magnolia Bakery on a Sunday morning and then catching a matinee showing of the recent *Chronicles of Narnia* movie—all set to hard-core rap stylings. It was goofy and catchy, and was also probably a throwaway segment on the show's first airing. But as fate would have it, shortly after the original broadcast, someone posted a video capture of the skit to YouTube, where it quickly racked up 5 million views.[21] NBC's lawyers had it taken down in a matter of days, but not before word of mouth around the video increased YouTube traffic by 83%.

After the early months of indifferent traffic, YouTube's audience exploded faster than any previous website in history (including Google, Myspace and Facebook). By the beginning of 2006, the site was serving 3 million video views a day. Six months later, that number had grown to 100 million views a day. Like most good Web 2.0 companies, YouTube achieved this success on a shockingly small amount of money. The company only ever raised $11.5 million, in two investment rounds. The fact that YouTube could serve video to the world from just a handful of servers (and some helpful content delivery networks in the background) was a powerful testament to the infrastructure the dot-com bubble had bequeathed to this new generation of startups.

Today we're used to popular "memes" bouncing around the world in an instant and have come to expect that social media can make superstars of teenagers from Canada (I'm thinking specifically of Justin Bieber, of course, who would be discovered thanks to videos his mother posted to YouTube). YouTube was ground zero for things like that, for the birth of modern meme culture as well as the social media–celebrity ecosystem. The idea that random events or random people could "go viral" really entered the mainstream thanks to YouTube. "We are providing a stage where everyone can participate and everyone can be seen," Hurley told the Associated Press in April of 2006.[22] There was no greater Web 2.0 manifesto than that.

But the "Lazy Sunday" phenomenon also pointed to one looming issue that concerned a lot of people about YouTube: there was a ton of copyrighted material uploaded illegally on the site. Sure, there were user-created home movies by the barrelful; but just as common were copies of last night's episode of *Survivor* or even clips from first-run movies still in theaters. In short, there was plenty of piracy going on. Just as with Napster, users came to expect that they could watch anything and everything on YouTube—from the latest Justin Timberlake video to obscure Japanese films from the 1960s.

But that was the issue: how was YouTube anything but Napster 2.0, with all the inevitable liability headaches that would imply? That was why people were obsessed with the who-will-buy-YouTube guessing game in 2006. Even though YouTube was exploding in popularity, it wasn't making any money, and in the postbubble era, an IPO was out of the question without meaningful revenue on the bottom line. So, unless YouTube was able to sell out to a deep-pocketed patron before the lawsuits started flying, it ran the very real risk of being pushed into an early grave.

As would come out in subsequent litigation, the YouTube guys knew perfectly well that there was a ton of pirated material on their site. But they had learned the lessons of Napster. Napster had attempted to make the argument that it enjoyed legal immunity under the Digital Millennium Copyright Act as a neutral platform. Service providers and platforms were protected as "safe harbors" under the law, provided they quickly and efficiently remove copyrighted material when notified. That was what had ultimately doomed Napster: it had never been able to take down 100% of the pirated files on its service. Five years on from Napster, might YouTube be able to find someone who could create a better system to remove illegally uploaded material—someone who had a mastery of algorithms, perhaps?

On October 9, 2006, Google announced that it was purchasing YouTube for $1.65 billion in stock. For the YouTube guys, selling to Google was logical: for all of YouTube's frugality, the cost of serving hundreds of millions of videos would eventually become prohibitive. Bandwidth might have been cheaper now, but who could hope to manage data on a scale that YouTube was achieving? Google was a perfect fit because its enormous infrastructure allowed YouTube the chance to handle the scale.

But Google's decision to take on YouTube's burden seemed downright crazy to a lot of people. Wasn't Google paying a lot of money to basically assume a huge liability risk? It turned out that Google made one simple calculation when it purchased YouTube: in the broadband era, video was likely to become as ubiquitous on the web as text and pictures had always been. YouTube was already, in essence, the world's largest search engine for video. In fact, it would eventually become the second-most-used search engine, period. With its stated mission to organize all the world's information, Google simply couldn't let video search fall outside its purview.

Google was able to come up with sophisticated automated systems that quickly and efficiently took down copyrighted videos when the rights holders alerted them. Lawsuits from aggrieved rights holders did eventually come, especially a billion-dollar lawsuit from Viacom. But because Google could prove that it was effective in policing content, in 2010 the judge in the Viacom case ruled in Google/YouTube's favor, saying that Google's takedown system was efficient enough that it complied with the Digital Millennium Copyright Act.

Google was the savior Napster never had. It had the infrastructure to allow YouTube to scale up; it had the technical sophistication to keep YouTube on the right side of the law; it had the money to contest the legal battles; and—most important—it provided YouTube with the business model that would allow it to thrive. Those little text ads that Google had put all over the Internet? They could be used to monetize the videos on YouTube just as they could with any other type of content. As the years went by, the text ads could even morph into actual video ads—but algorithmically targeted and effective ads, as Google's ads always were.

And this was the last way in which YouTube's timing was impeccable. The movie and television studios had watched the Napster debacle with dread. They knew their industries were next in line for disruption from the Internet. When that disruption arrived, in the form of YouTube, Hollywood was at least willing to weigh its options this time. Going scorched earth against Napster had not saved the music industry. And so, once Google came to the table with a willingness to share advertising revenue with rights holders, a lot of them (Viacom notwithstanding) were willing to play ball. At least Google/YouTube was offering Hollywood *some* kind of revenue stream. Digital revenue might not be as lucrative as the old analog revenue streams but, well, that was the Napster lesson, right? Better to take what you could get and embrace new distribution models rather than fight them. The entertainment industry was even now willing to buy into one of the key arguments Napster had tried to make only half a decade before: giving users a taste of your content online was actually great promotion! The phenomenon of *Lazy Sunday* had shown that. By 2008, when YouTube was streaming 4.3 billion videos *per month* (in the United States alone), many people—young people especially—were beginning to watch more video online than they were watching on traditional TV.[23] For the first time, Hollywood stopped fighting disruption, and followed the changing tastes of their audience into a digital future.

■

WEB 2.0 WAS ABOUT PEOPLE expressing themselves—actually being themselves, actually living—online. The last piece of the puzzle was simply to make the threads of all this social activity explicit.

Online chat clients like IRC, through which the Napster hackers had met each other and collaborated, had a technological cousin at AOL. In the days when AOL was still the dominant ISP with more than 20 million users, its internal messaging program allowed you to chat with your friends and family in real time. AOL's chat had an extra feature called the "Buddy List" that alerted you as to which of your friends were online at the same time you were, so you could hit them up for a quick conversation. The system also allowed you to leave an away message so that your friends could know when they might expect you to be online again.

Instant messaging was only intended for internal use by members of AOL's walled garden. But in 1997, the company did something completely out of character: it released the messenger program online as a stand-alone web client. It was known as AOL Instant Messenger, or AIM, and it allowed people to stay in touch with their AOL friends when they were away from AOL. It proved especially popular for people who were at work, where they couldn't log on to AOL, and among teenagers, allowing them to keep up with all of their friends, whether they were AOL users or not. Soon, there were hundreds of millions of AIM users, many times more than the number of actual AOL subscribers at its height. Even as AOL the company began to crumble after the disastrous merger with Time Warner, AIM continued as a breakout success for one simple factor: it was a literal social graph, a tangible map of your online connections and relationships. Chatting on AIM became more popular than email, and your AIM screen name eventually gave you the ability to customize a rudimentary profile, turning it into a valuable online marker of identity. These features, combined with the away messages and status updates, came to reflect a user's daily circumstance. Add to this the emojis and icons that allowed AIM users to project their mood, and AIM became a fully functional and real-time representation of the digital self. There was even an abortive project to create "Aimster," which would add the ability to search a friend's hard drive and trade files (AOL management, of course, killed that before it could see the light of day).

And that was the problem, of course. AOL had no idea what it was sitting on. AIM was a fully fleshed-out social network. True, it was free to use; but it was making a limited amount of money thanks to traditional banner ads. Had anyone at AOL been able to predict the future, AIM could have been the perfect platform to transition AOL users into the post–dial-up world. Before we were all sending SMS texts, before we all reconnected on Facebook, a great many of us were connected on AIM. The social graph was actually *the* great prize of Web 2.0. Others were only able to seize this prize because AOL dropped the ball. AIM eventually lost its relevance through benign neglect. "If AOL had 20/20 hindsight, maybe the story [of social networking] would have had a different ending," says Barry Appelman, one of the AOL engineers who invented AIM.[24]

■

SOCIAL NETWORKING MIGHT SEEM like a dead-obvious concept in retrospect, but that's only because we've gone through the looking glass into a modern world where the boundaries between our online lives and "real life" have been broken down almost completely. The roots of social networking go all the way back to the early web. The earliest dating sites like Match.com and the message boards on sites like iVillage allowed users to create an online "profile" or representation of your real-world self. And sites like GeoCities and Angelfire allowed users to construct personal webpages so intricate as to serve as virtual avatars in cyberspace.

The first modern social-networking site as we would recognize it today was invented by SixDegrees.com. In 1996, a former lawyer and Wall Street analyst named Andrew Weinreich had an idea inspired by the popular notion that any single person on the planet can be connected to anyone else by around six steps of personal connections—"six degrees" of separation. If that was true, then the web was the perfect tool for mapping those connections.

Launched in early 1997, SixDegrees took off in about a month, in the usual viral way we're now familiar with: users sent their friends invitations to link up on the site. At its peak, the site had 3.5 million members, and in 1999, Weinreich wisely sold the company for $125 million to another Internet startup.[25]

At the time, many viewed SixDegrees as a newfangled Rolodex at best, a creepy dating site at worst. But Weinreich had been convinced there was something more powerful to the idea of networking online. "We envisioned Six Degrees being something of an OS—of an operating system—and we thought about it in the context of when you're buying a watch at eBay you should be able to filter the watches based on people's proximity to you," Weinreich said. "You should be able to filter movie reviews in the future by who's reviewing them."[26] It was the right idea, but as Weinreich would ruefully admit, "We were early. Timing is everything."[27] The site was expensive to operate in the dot-com days, and of course, there were no photos on the profiles. "We had board meetings where we would discuss how to get people to send in their pictures and scan them in," Weinreich says.[28] After the dot-com crash, the site was shuttered.

In 2002, a former Netscape employee named Jonathan Abrams launched a site called Friendster. Abrams wanted to rekindle SixDegrees' original notion of real identities and real personal connections. Within a few months, the site had 3 million users from word-of-mouth marketing alone.[29] The media seized upon Friendster as a more sophisticated version of online dating, and certainly, the digital profile pictures that could now easily be uploaded to your Friendster helped shape this impression of the site. Once connected to someone else, you could browse their friends to see who among them was attractive (and single) and then the idea was that your friend would put in a good word for you. But, this was just as the notion of the Web 2.0 renaissance was taking hold in Silicon Valley, so, dating site or no, Friendster was able to raise $12 million from blue-chip VCs including Kleiner Perkins and Benchmark Capital. In 2003, Google offered to buy Friendster for $30 million in pre-IPO Google stock, but the venture capitalists encouraged Abrams to spurn the offer and instead shoot for the moon.

Friendster ended up missing the moon by some distance. It turned out that hosting blogs or even serving portal pages to millions of users was one thing, but a social network scaled to millions of users was another thing entirely. On a social network, the content was ever-changing, and what was served to each user was often unique to that user, often only in that moment of time. Friendster had to dynamically propagate each new update, each new post—and each new picture. The engineering challenges of delivering what was quickly becoming a deluge of content were at a whole new scale, and Friendster simply wasn't up to the challenge.

"When it grew as fast as it did, we absolutely weren't prepared for it," Abrams said later. "Throughout 2004, 2005, Friendster barely worked. The site was really slow; it was buggy. That, unsurprisingly, caused an exodus of users to leave."[30] When Friendster users grew frustrated waiting thirty or more seconds for pages to load, they had a throng of Friendster copycat sites to turn to instead. Like any good idea, the rebirth of social networks inspired dozens of people to try their hand at the concept. Many of the Friendster copycats tried to create social networks that targeted specific niches: college students, high school students, even, in the case of Dogster.com, pet owners.

One of the copycat sites that rushed in to tempt away disillusioned Friendster users was called Myspace. Myspace was owned by eUniverse, a dot-com survivor that made a lot of money peddling wrinkle cream ("Better than Botox") via online ads that purported to offer the cream for free despite built-in expensive automatic refills, and that made advertising claims that the FDA asserted "were not supported by reliable scientific evidence." An eUniverse employee named Tom Anderson became obsessed with Friendster and convinced his boss, Chris DeWolfe, that creating a Friendster clone might be a cheap and easy way to amass more people for eUniverse's marketing lists. On August 15, 2003, Myspace was launched as a nearly feature-for-feature clone of Friendster. Users had a profile page where they could post pictures, share their interests and hobbies, and link to the profiles of their friends and family. But Myspace also added a kitchen sink's worth of features, such as blogs, horoscopes, games and more.

One of the things that was driving users away from Friendster (aside from the slow performance) was the fact that Abrams had insisted on a strict fidelity to identity. Anytime users created a Friendster account under a pseudonym, or started a parody account or pretended to create an account as a celebrity, Friendster would delete it. Myspace had no such regulations. If you wanted to sign up as Leonardo DiCaprio or Bugs Bunny, Myspace let you do it. Furthermore, you could follow anyone you wanted, whether you truly knew them or not. Myspace was the first to hit on a key concept in social networking: linking to others could be a way of mapping your personal connections, but it could also highlight your personal tastes. Friending, or "following" another profile, could be a powerful vote of interest and engagement. When this was combined with the ability to host MP3 files on your profile, Myspace became a potent venue for promotion, especially among musicians. Now that Napster was gone, an entire generation of unknown musical acts ranging from Fall Out Boy and My Chemical Romance to Arctic Monkeys would rise to prominence by engaging with their thousands of fans, promoting tour dates and even releasing new songs on their Myspace pages.

Myspace also had a laissez-faire attitude when it came to self-expression. Users could redesign their pages at will, hacking into the design code itself to create flashy, colorful, even garish profiles. This appealed especially to teenagers, who decorated their Myspace pages like they would decorate the walls of their adolescent bedrooms. Myspace also looked the other way when users posted racier content. Profiles featuring scantily clad women abounded. This side of Myspace was exemplified by Tila Tequila, a young Vietnamese-American model who was one of the many users fed up with Friendster. "I was getting too many friend requests, and the pictures were too hot," Tequila said about Friendster's habit of repeatedly banning her profile.[31] So she took the tens of thousands in her digital audience to Myspace, where she could represent herself however she wanted. Soon her "friends" numbered in the hundreds of thousands and Tequila achieved that unique mid-2000s form of D-level fame. "There's a million hot naked chicks on the Internet," Tequila told *Time*. "There's a difference between those girls and me: Those chicks don't talk back to you."[32]

Thanks to all of these factors, Myspace quickly rocketed past Friendster to become the king of the social networks, racking up 1 million users less than six months after launching and 3.3 million after a year of operation, with 23,000 new users signing up daily.[33] By May of 2005, Myspace was attracting 15.6 million visitors every month.[34] Myspace founders Tom Anderson and Chris DeWolfe became celebrities in their own right. In Anderson's case, it was because he was the guy who interacted with the users; by default, Tom was every new user's first friend. For his part, DeWolfe put himself forward as Myspace's strategic visionary. "We want to be the MTV of the Internet," Wolfe told investors.[35] To the *New Yorker*, he proclaimed: "The Internet generation has grown up, and there are just a lot more people who are comfortable putting their lives online, conversing on the Internet, and writing blogs. This generation grew up with Napster and the iPod."[36] Myspace was just serving this new audience's behavior and expectations.

But the story of Myspace is slightly different from that of the other companies in the Web 2.0 wave. For one thing, Myspace was Los Angeles–based, a key factor that may have contributed to the site's focus on glam and glitter. And—uniquely—Myspace wasn't a startup. Rather, it was a subsidiary of a parent company. Anderson and DeWolfe weren't actually calling the shots at Myspace. That parent company, eUniverse, had rebranded itself as Intermix in order to escape the shadow of its seedy past, and as the excitement over Web 2.0 grew more frenzied, Intermix decided the time was right to cash in on Myspace. In July of 2005, Intermix announced that it (and therefore, Myspace) had been acquired for $580 million. The acquiring party was not a Google, or even a Yahoo, but News Corp, the company run by media mogul Rupert Murdoch.

Coming as it did among the slate of other Web 2.0 acquisitions, as soon as the deal was announced, many in the press and even some in the tech industry itself were quick to announce that another bubble had formed in Silicon Valley. But for a while, Myspace's unbelievable growth made those fears seem far-fetched. By the end of 2005, a mere six months after the acquisition, Myspace could claim about 40 million registered users and more monthly pageviews than eBay, AOL or even Google.[37] By the time Myspace inked a $900 million advertising partnership with Google in 2006, it looked like social networking was, indeed, the next big thing. MySpace was the new 800-pound gorilla on the web, and Rupert Murdoch had pulled off the steal of the new digital century.

But even when Myspace was at its zenith in terms of users and traffic and revenue, people couldn't stop comparing it to another of the Friendster clones, particularly the clone that had chosen to focus exclusively on college students. In a November 2007 News Corp earnings conference call, Rupert Murdoch himself dismissed this competitor, Facebook, as merely a "Web utility similar to a phone book." Myspace, by comparison, had "become so much more than a social network. It connects people, but it's evolved into a place where people are living their lives. A social platform packed with search, video, music, telephony, games."[38] Little did Murdoch know that, even as he said those words, the battle for social networking was already over, and Myspace would join SixDegrees and Friendster as an also-ran in the history books.

15

THE SOCIAL NETWORK

Facebook

It's something of a universal phenomenon that we can probably all recognize from our own lives. When you're between the ages of sixteen and twenty-four, you're plugged into the zeitgeist. During that intellectually fecund period, you tend just to "get" things: the latest fashions, the coolest new music and films, the trends and jokes and ideas that are au courant. It's almost like young people see the future before everyone else.

Mark Zuckerberg was eleven when Netscape IPOed. As a middle-schooler and high-schooler, he came of age on AOL. In 1999, he had a personal homepage on Angelfire, a competitor to GeoCities where anyone could host a website for free. "Hi, my name is . . . Slim Shady," the site's About Me page said.

> No, really, my name is Slim Shady. Just kidding, my name is Mark Zuckerberg (for those of you that don't know me) and I live in a small town near the massive city of New York. I am currently 15 years old and I just finished freshman year in high school.

A subpage on young Mark's website called "The Web" had a Java applet on it that plotted out a graph illustrating connections between people Mark knew. He asked his friends to link to each other on the applet, so he could plot out his teenage social circle.

It would be poetic to think that here, in 1999, was the germ of an idea that would later become Facebook. But the truth is, Mark Zuckerberg was just plugged in to the web's zeitgeist. Sharing, connections, social media, these were all impulses bubbling to the surface and, at fifteen years old, as a web- and computer-obsessed kid, Mark Zuckerberg sensed these trends intuitively. Zuckerberg, like almost everyone he knew, was a heavy AIM user. He was also a member of Friendster when it debuted. He blogged. He voted on HotorNot.com. Napster had been the biggest cultural and technological event of his young life. And so, Zuckerberg's youthful hacks all featured elements that, in one way or another, we might call "social."

As a senior in high school, Mark and fellow student Adam D'Angelo developed Synapse, a clever plugin to Justin Frankel's Winamp that sampled the MP3s a person listened to and then algorithmically generated playlists based on that user's taste.[1] D'Angelo had previously created Buddy Zoo, a program that, much like Zuckerberg's "The Web" applet, made a graph of your personal connections—but in this case, using AIM. The boys received buyout offers from Microsoft and AOL. The pair opted for college instead.

Zuckerberg enrolled in Harvard, to major in psychology. But even matriculation at one of the world's most prestigious schools didn't hamper Mark's penchant for hacking. During his sophomore year, Zuckerberg created an online app called Course Match that helped his fellow Harvard students choose what classes to sign up for, based on who else was signed up for that class already. That way you could rub shoulders with your friends, or maybe that cute girl you wanted to meet. Later that same year, when Zuckerberg got behind on coursework for a class called "Art in the Time of Augustus," he put up a website encouraging his classmates to contribute to a collective analysis of the artworks in question, Wikipedia-style. This clever gambit allowed Mark to quickly cram and pass the exam.[2]

Zuckerberg also created a HotorNot for Harvard students called Facemash that let users vote on the looks of their fellow classmates. "Were we let in [to Harvard] for our looks?" the site asked. "No.

Will we be judged by them? Yes."[3] Facemash was an instant hit on campus, but was quickly shut down because Zuckerberg had stolen the student profile pictures used on the website from Harvard's internal networks. Also, student groups objected to the blatant misogyny and privacy violations inherent in the project. Zuckerberg was put on probation by Harvard's administration for the stunt.

"I had this hobby of just building these little projects," Zuckerberg would later say of his early programming endeavors.[4] Again, Zuckerberg was not unique in this—not some lone genius churning out social apps because he had some singular or unprecedented insight. Rather, he was part of the web's collective unconscious, groping blindly toward what would soon be known as Web 2.0. Zuckerberg wasn't even unique in pursuing social apps at Harvard! After the controversy surrounding Facemash gave him a brief bout of campus celebrity, Zuckerberg was contacted by a trio of Harvard students, Divya Narendra and identical twin brothers Cameron and Tyler Winklevoss, who were working on a college-based social network they wanted to call HarvardConnection.

Sharing. Social networks. Mapping relationships online. It was just in the air at the moment. In the zeitgeist.

■

ZUCKERBERG AGREED TO HELP program HarvardConnection shortly before the winter break of the 2003–4 academic year. Sometime over that hiatus, it seems that he decided to abandon this project and instead take a crack at coding up a fully formed social network himself. Harvard had a decades-long tradition of publishing "facebooks," or directories of student portraits that helped people look each other up and make connections. The university had made some noises about bringing these directories online, and just that December of 2003, Harvard's student newspaper, the *Crimson*, had published an editorial titled "Put Online a Happy Face: Electronic Facebook for the Entire College Should Be Both Helpful and Entertaining for All."[5] Mark had already had experience with an online facebook. In high school, his classmate Kristopher Tillery had created a website that basically replaced the printed directory the school previously used.[6] It seemed silly that a high school could do an online facebook and Harvard couldn't.

Zuckerberg decided not to wait for the university to get its act together. On January 11, 2004, he registered the domain Thefacebook.com for $35. Using the examples of Friendster, Course Match, Facemash—even drawing from AIM, Buddy Zoo and Zuckerberg's own "The Web" app—Zuckerberg coded up a website that would bring college facebooks into the web era. He paid $85 a month for hosting to a company called Manage.com, and on Wednesday, February 4, 2004, he put the website live, along with the following message:

> Thefacebook is an online directory that connects people through social networks at colleges. We have opened up Thefacebook for popular consumption at Harvard University. You can use Thefacebook to: Search for people at your school; Find out who are in your classes; Look up your friends' friends; see a visualization of your social network.[7]

After putting the site live, Zuckerberg went out for pizza with his roommates. They discussed the Thefacebook project and how someday somebody was going to build a community site just like it—but for the whole world. Whoever pulled that off would create one amazing company. They wondered who would eventually do it. "But it clearly wasn't going to be us," Zuckerberg would recall later. "I mean, it wasn't even an option that we considered it might be us."[8]

Four days later, more than 650 students had registered as users of Thefacebook. By the end of the month, three-fourths of Harvard's student body was using the site daily.[9]

■

WHEN MARC ANDREESSEN STARTED Mosaic, he turned to his fellow students to help; when Shawn Fanning started Napster, he turned to his fellow hackers. Right away, as Thefacebook took off at Harvard,

Mark turned to his fellow dorm mates in Suite H33 of Kirkland, the undergraduate residential house he lived in, to keep the project afloat. Roommate Dustin Moskovitz was enlisted to help code the site and expand it. Suitemate Chris Hughes was recruited to help with promotion and serve as the site's spokesperson. A fraternity brother of Mark's from Alpha Epsilon Pi, Eduardo Saverin, was brought on board as a full business partner and to run the finances. Later, Zuckerberg would even turn to his old friend Adam D'Angelo (then at Caltech) to help Moskovitz with the coding.

Thefacebook was founded by a bunch of kids who had lived through the dot-com era as well as the Napster supernova. To them, starting a website—or even a web company—was not some crazy notion. On the contrary, it was aspirational, but also feasible. It was like starting a band or a student group, or maybe opening an off-campus bar. The Mosaic kids had been academic researchers who didn't know the first thing about startups. The Napster kids had been naïve hackers who didn't know the first thing about business or the law. But Thefacebook was started at Harvard, by the scions of America's elite families. These were the kids who were *supposed* to conquer the world in some way or another. So, when they got an idea for a cool website, they knew what to do: see how big it could get. And they had the resources to make that happen.

This is where the myth of Facebook's founding is colored by the artistic license taken by the movie *The Social Network* (and the book that inspired it, *The Accidental Billionaires: The Founding of Facebook*). Sure, Mark Zuckerberg was a bit socially awkward, but according to friends at the time, he didn't have much trouble getting girlfriends. He was confident. He was a leader. Zuckerberg and Saverin weren't old enough to drink, but they were familiar with money since they both came from privileged backgrounds. Zuckerberg had gone to the exclusive boarding school Phillips Exeter Academy. Saverin came from a long line of international businessmen. So, this wasn't a case of dorky social outcasts coding up a website in order to meet girls. Thefacebook was just a cool thing they could make together. If it ended up being a real company in the end (or, actually help them meet girls), well, even better.

Zuckerberg and Saverin both invested $1,000 of their own money into the project, and Saverin created an LLC and opened a bank account. Within weeks of launching at Harvard, Moskovitz and Zuckerberg began cloning the site and seeding it to other campuses. First came Columbia and Yale, and then Stanford. Dartmouth and Cornell followed the next month. After that: MIT, University of Pennsylvania, Princeton, Brown and Boston University. The uptake at each new school was just as instantaneous as it had been at Harvard. By the end of March 2004, Thefacebook had 30,000 users.[10]

This was viral growth, but, crucially, it was *managed* viral growth. By expanding to colleges one at a time, the five founders could grow the site without the rate of growth outpacing them. The boys had learned by watching Napster, and especially by watching as the Friendster fiasco unfolded before their eyes. They only released Thefacebook to a new college when they knew they had the infrastructure in place to handle the additional traffic. They studiously avoided site crashes and service outages. They made sure the pages loaded quickly by assigning each school to a unique database, thereby avoiding the complicated networking calculations that slowed down Friendster.[11]

This staggered growth also allowed the company to expand within its financial means. In true Web 2.0 fashion, Thefacebook was run frugally, using free open-source software like MySQL for the database and Apache for the web servers. Even by the time users were in the tens of thousands and Thefacebook was live on dozens of campuses, it was only costing $450 a month to run the site off of five Manage.com servers.[12]

Thefacebook focused on colleges because that is what its founders knew. As Sean Parker would later say of the embryonic company, Zuckerberg wanted Facebook to get big. "But he didn't know what that meant. He was a college student. Taking over the world meant taking over college."[13] Whether by accident or design, the self-enforced exclusivity of focusing on colleges was key to Thefacebook's early success. We are never more social than we are in college; our network of friends and connections is never more vibrant and vital than in those years. From day one, Zuckerberg's vision for Thefacebook mimicked the original instincts of SixDegrees and the best intentions of Friendster. You could only register on Thefacebook with your college-supplied .edu email address. You could only interact with other students at your actual school. You had to be your authentic self, just as you would on campus. There would be no fakesters or parody accounts on Thefacebook. But then, no one would want those anyway. Being inauthentically yourself was to miss the point of Thefacebook entirely.

Thefacebook attempted not merely to re-create your offline social circles, or to build new types of social connections online. No, Thefacebook wanted to mirror your *exact* social circle. The friends you took classes with, the friends you roomed with, the friends you sat in the dining hall with—those were your friends on Thefacebook. Mapping your social network accurately on Thefacebook provided a new, frictionless way to map your social world—to curate it, to *live* it. Thefacebook understood that you didn't need any bells and whistles to make a social network compelling. If users were willing to port their actual social lives onto the Thefacebook's network, then the network could be as compelling and vital as offline life was.

As an early user (who would go on to be an early Facebook employee), Katherine Losse would write, describing her first encounter with Facebook as a student at Johns Hopkins University in 2004:

> It was the first Internet site I had ever used that mirrored a real-life community. The cliques on Facebook were the same ones I ran into at the library and campus bar, and the things people said to each other on their walls—water polo team slang, hints at the past weekend's conquests, jabs at Hopkins' lacrosse archrival Duke—were similar to what you heard them saying at study tables or around pitchers of beer. The virtual space mapped the human space.[14]

By targeting the narrow field of your actual college social circle, Thefacebook was able to construct a digital social web that directly paralleled reality. It was a living online Rolodex, just like Andrew Weinreich had tried to achieve in theory; it was your true self, projected virtually, as Jonathan Abrams had aspired to, and failed. Thefacebook actually achieved true digital identity.

Choosing to launch only at elite colleges also helped. Thefacebook had an air of exclusivity. It was the social network (at least at first) of the elite, the 1%. It further helped that Thefacebook felt classier than Myspace, which was exploding in popularity at the same time. The aesthetic of Thefacebook was almost the antithesis of Myspace; it was less flashy, more functional; presentational where Myspace was promotional. You didn't go to Thefacebook to show off, but you did go there to present your best self.

Thefacebook's early features were purposefully limited. You could map your connections to your fellow students, and originally, your connections were restricted to your own school. As Thefacebook started spreading to other colleges, you could eventually connect to outside friends only if you both mutually acknowledged your relationship. You could only post one photo: your profile picture. You could fill out a whole range of personal categories ranging from sex and relationship status to courses enrolled in, extracurricular activities, hobbies, favorite films and the like. There was a "status update" feature taken directly from AIM. And there was the ability to "poke" other users, which meant—well, no one was exactly sure. But it was college, so if you poked someone, it could mean whatever you wanted it to mean.

It's important to see Thefacebook for what it was at this moment: a social directory. A cool little utility. Zuckerberg himself repeatedly described the project as a "social utility." This was just another one of his hacks, but it happened to be one that had gotten the most traction. It was no different than Course Match or anything else he had done previously. In fact, had Facemash not been shut down by the authorities, perhaps Zuckerberg would have ridden *that* wave instead of this one. Maybe he would have enlisted his friends in building HotorNot-like sites for Stanford and for Yale and on and on. In fact, on his blog, right around the time he released Facemash, Zuckerberg had suggested exactly that: "Perhaps Harvard will squelch it [Facemash] for legal reasons without realizing its value as a venture that could possibly be expanded to other schools (maybe even ones with good-looking people . . .)."[15]

Well, Harvard hadn't squelched Thefacebook, and the boys in Kirkland House were going to take it as far as they could. But that required more money, of course. So, from very early on, Facebook had ads. That was Saverin's main contribution to the project. He was indeed business savvy and he did have actual connections to advertisers. Saverin hooked Thefacebook up with Y2M, a company that sold ads for college newspaper websites. Cannily pitching Thefacebook as a new way to reach the coveted college demographic, Y2M began brokering ads on the site. One of the first advertisers was MasterCard. Unsure of Thefacebook's viability as a marketing tool, MasterCard refused to pay up front, or even to pay for pageviews served. They were willing only to pay a flat fee if a user actually opened a new credit card account. Within a day of launching the ads on Thefacebook, there were twice the applications MasterCard had anticipated for the entire four-month campaign that had been planned.[16]

Saverin continued shaking the trees and landing deals like this, depositing the proceeds into the bank account he controlled. He and Zuckerberg both invested $10,000 more of their own money to serve as working capital. But almost from the first weeks, Saverin was also lining up meetings with financiers. At one meeting that June, an investor offered $10 million for the company, which was barely four months old. And at another meeting, in New York City in April, Saverin and Zuckerberg met Sean Parker (of Napster fame) for dinner. The famous line from *The Social Network* movie is "A million dollars isn't cool, you know what's cool? A billion dollars." That's just dialogue invented by the screenwriter, Aaron Sorkin, but what is true is that the dinner depicted in the movie really did take place (at Jean-Georges Vongerichten's 66 restaurant in Tribeca) and Zuckerberg really was awed by Parker's geek celebrity.

And that dinner does seem to have been a turning point in Thefacebook's destiny. From twenty-year-old Mark Zuckerberg's perspective, it felt like maybe he was sitting on some sort of web phenomenon. Perhaps he could be the next Shawn Fanning or Sean Parker. Napster, of course, was a cautionary tale, a tragic failure. But maybe Zuckerberg could do better. He wanted to give it the old college try. And that, he decided, meant leaving college (temporarily, at least) and heading out to California. Where the Internet happened. And so, when the spring 2004 semester wound down, he rented a house in Palo Alto and moved out for the summer, along with Dustin Moskovitz and three other Harvard friends/interns.

At that point, Thefacebook had launched at thirty-four schools and had 100,000 users.[17]

■

WE'VE SEEN HOW the startup culture of modern Silicon Valley was created to serve the habits and metabolism of postcollege white males, especially (although in slightly different ways) in the examples of Netscape and Google. But the summer of 2004 that Thefacebook spent in a rented ranch house on a cul-de-sac at 819 La Jennifer Way has gone down in lore—at least in some circles—as the brotastic, edenic ideal of an Internet startup's incubation. These weren't college graduates, these were college *sophomores*. So: There was a swimming pool. There was a jury-rigged zipline that was strung from the chimney so that you could drop down from the roof into the pool. There was alcohol and marijuana at all hours. There were beer-pong tournaments. There were parties. These were the guys who were running the most popular college-based website in the world, after all, so when the boys wanted to throw a kegger, they just posted a notice on Thefacebook pages of nearby Stanford University. Hundreds of kids would show up. People passed out on the floor and slept where they landed. Friends and hangers-on would come and crash on the couch, sometimes staying for weeks. The whole house was littered with used soda cans and empty pizza boxes.

But amid all of this, on desks and in corners and sometimes out by the pool, there were kids hunched over their keyboards coding up one of the hottest websites in the world. It was a startup, but it was a startup in the hands of nineteen- and twenty-year-olds: just as much frat party as work. Zuckerberg himself usually didn't start programming until the early afternoon, but the coding sessions could stretch on until dawn—in spite of whatever other activities were going on in the house. Even if there was loud music playing or a raucous party going on in the background, everyone working on Thefacebook tended to communicate over AIM anyway, even when they were sitting right next to each other. Noise was not an issue. Distraction was not a factor. All summer, at all hours, there was almost always somebody, head down, staring at lines of code on a computer screen.

"We were doing fourteen- or sixteen-hour days," Moskovitz recalled later. They mostly worked in the kitchen on their personal computers and, in Moskovitz's words, "hammered away."[18] The goal that summer was to prepare for classes to resume in the fall. The expectation was that, come September, Thefacebook would launch on seventy new campuses.[19] There were new features to test, new servers to bring online. But at the same time, there was still a sense that this was all some elaborate (but "kind of" serious) lark. When a reporter from the *Crimson* stopped by to check in on these wayward Harvard boys, Zuckerberg described the operation this way: "Most businesses aren't like a bunch of kids living in a house, doing whatever they want, not waking up at a normal time, not going into an office, hiring people by, like, bringing them into your house and letting them chill with you for a while and party with you and smoke with you."[20]

It was just kids playing grown-up, seeing how far they could take things. Whether it was posturing or not, the official line was that they'd all be heading back to Harvard in the fall to continue their studies. "We like school and want to go back to school and at some point somebody's gonna offer us a lot of money and we'll probably take it, you know?" Zuckerberg told the *Crimson*.[21] Until that happened, they were just living the Silicon Valley startup fantasy. Zuckerberg even seemed to be hedging his bets, concentrating a lot of his time on a Napster-like file-sharing program called Wirehog, which he intended to integrate into Thefacebook's feature set so that users could trade MP3s, videos, files, what have you. It seems that, despite

Thefacebook's success, even Mark Zuckerberg wasn't exactly sure that this social-networking thing was much more than that almost dismissive word he used to describe it: a utility.

And then into this scene came Sean Parker.

If there was anyone who was plugged into—who virtually embodied—the web's zeitgeist, it was Sean Parker. What had fascinated him the most during his time at Napster were the social elements of the thing. The sharing. He wasn't surprised when these trends resurfaced in Napster's wake. After he was pushed out of Napster, he founded a new startup called Plaxo, which used people's email and contact lists to, almost literally, put everyone's Rolodex online where it would be searchable, shareable and constantly updated. It was a virtual white pages of everyone's contact info. Parker was convinced that mapping digital identity was the next big thing. And in Thefacebook, he saw the purest expression of this idea so far.

Parker was the one who had initiated that New York dinner with Zuckerberg after watching Thefacebook take over Stanford's campus, where his then-girlfriend was matriculating. Now that Thefacebook was, at least temporarily, carpetbagging in his Silicon Valley stomping grounds, when he and Zuckerberg crossed paths in Palo Alto (that same girlfriend lived down the street from Thefacebook house), Sean Parker jumped on board as Thefacebook's most committed true believer.

In fact, he moved into the house.

Parker, like everyone else involved in Napster, had not made very much money when the company went belly-up. And even though Plaxo was enjoying some measure of success, Parker was, at that very moment, in the process of being pushed out of his latest startup as well. But now the issue was not careless emails. Now the whispered accusations were about partying, drugs, and generally erratic behavior. Whether those accusations were true, or whether they were just part of a smear campaign, as Parker claimed, when Parker moved into 819 La Jennifer Way, he was not only between gigs, he was quasi-homeless.

But Mark Zuckerberg continued to hold Parker in great esteem. Everyone in the house did. Parker was five years older, for one thing, so he was of age and could keep the house well stocked with alcohol. And he had a car. The boys from Thefacebook had simply been walking everywhere. Most important, Parker had already played an integral role in the launch of two major web startups. To Zuckerberg and Thefacebook team, he was basically a grizzled Silicon Valley veteran. As Zuckerberg spent the summer considering his options, and considering the possibilities for Thefacebook going forward, he increasingly turned to Sean Parker for counsel. "You trust people you can relate to; I could relate to Sean," Zuckerberg would say later. "And I was impressed he had done something cool."[22]

Zuckerberg would later say that he and Parker bounced so many different scenarios off each other that summer that he's not sure, in retrospect, which ideas were Sean's and which ideas were his. But if there was one idea Parker seemed hell-bent on drilling into Zuckerberg's head, it was that Thefacebook was *the thing*. Zuck should just stick to his instincts and keep with the original game plan: build out Thefacebook school by school and see how big it could get.

"I've really got something here?" Zuckerberg asked one evening.

"Yeah, Zuck, you do," Parker said.[23]

At Parker's urging, Zuckerberg decided that Thefacebook shouldn't just plan for the immediate future; it should plan for an exponential future. To prepare for the coming autumn and the anticipated influx of users, Thefacebook desperately needed new servers. Zuckerberg decreed that, rather than struggle to keep up, the site's infrastructure should, from that point forward, be architected to anticipate ten times the number of users it was getting at any one moment. That would cost more money than Facebook was already generating. Zuckerberg and his family were forced to sink $85,000 into the company, mostly for buying new servers.[24]

The time had clearly come to land serious VC backing. But Zuckerberg and the others had listened in that summer as the humiliating legal process of Parker's ouster by Plaxo's investors played out to its sorry conclusion. The experience gave Zuckerberg a sobering education about what he might be in for ("VCs sound scary," he remembers thinking).[25] So, when it came time to shake the trees for money, Sean Parker made it his mission in life to make sure Thefacebook got a good deal.

Parker introduced Zuckerberg to LinkedIn's founder Reid Hoffman, as well as Mark Pincus, a Web 2.0 entrepreneur who had founded another early social network, Tribe.net. Both made angel investments in Thefacebook. Parker also got Zuckerberg a meeting with the de facto head of the PayPal Mafia, Peter Thiel. Thiel gave Zuckerberg a $500,000 loan, which would convert into about 10% of the company's equity. The terms were generous, and Parker was confident that Thiel was the sort of investor who would leave Zuckerberg alone to pursue his vision. The only instruction Thiel gave the twenty-year-old was: "Just don't fuck it up."[26]

Thiel did ask if the boys were still planning on returning to Harvard in the fall. Zuckerberg said yes.

"Okay," Thiel said. "*Sure* you are."[27]

By going the angel route and avoiding big-name venture capital firms, Parker ensured Zuckerberg maintained majority control of the company's precious equity. Parker also reincorporated Thefacebook as a proper company, jettisoning the old LLC structure set up by Saverin, and further consolidated Zuckerberg's control (Parker also gave himself a healthy chunk of equity and a seat on the company's board of directors for his troubles). With this cash infusion, Thefacebook would have the funds necessary to meet the expected fall crush head-on. And Zuckerberg would control where the company went from there.

Which was a good thing, because Zuckerberg wasn't going anywhere. It turned out that Peter Thiel had sized up the boys of Thefacebook correctly. When the summer ended and the crucial fall season approached, Moskovitz and some of the others agreed to take a semester off, stay in California, and see how things went with the major school expansion. The idea of returning to school seemingly faded into the background after that, never to be seriously considered again.

■

IN THE FALL OF 2004, Thefacebook went gangbusters. Even though it was supposed to be a slow period, the user base had actually doubled over the summer, to 200,000.[28] In September alone, that number doubled again as new schools were brought online.[29] The site also rolled out two major new features. Each profile now had a "wall," which was like a virtual corkboard outside a dorm room—a place where you or your friends could post messages and greetings. And now there were also ad hoc "groups" that you could join, for things like study sessions and campus causes, but really, anything under the sun.

On November 30, 2004, Thefacebook passed the million-user mark. It had been live for all of ten months.[30]

And yet, Zuckerberg still did not seem convinced that Thefacebook was his meal ticket. "What was so bizarre about the way Facebook was unfolding at that point," Sean Parker has said, "is that Mark just didn't totally believe in it and wanted to go and do all these other things."[31] The main "other thing" was Wirehog, which was taking up just as much of Zuckerberg's time—if not more. There was also the continued sense of kids-playing-dress-up. Zuckerberg had business cards printed up that read: "I'm CEO . . . bitch!" It was probably a riff on the then-ubiquitous Rick James sketch from *Chappelle's Show*, but as early Facebook employee Andrew "Boz" Bosworth has written, the card also spoke to "how unclear it was even in his own mind at the time that he would someday become such an important (and scrutinized) leader."[32]

It was around this time that Zuckerberg infamously showed up late to a meeting with the venture firm Sequoia Capital, still dressed in pajamas and pitching from a PowerPoint presentation that included a slide with the title "The Top Ten Reasons You Should Not Invest."[33] This incident was a prank instigated by Parker, who had a grudge against Sequoia, blaming them for his exile from Plaxo. Any entrepreneur who was even halfway serious about his reputation in Silicon Valley would never be so openly contemptuous of one of the most successful VC firms in the tech universe. Zuckerberg later apologized for the stunt.

Three things conspired to turn Zuckerberg's attitude around and get him to take Thefacebook seriously. First, Wirehog was a dud. After it was launched on Thefacebook in November of 2004, essentially nobody used it. So, Zuckerberg's notion that social media was more important than social networking was proven wrong.[34] The second factor was competition, pure and simple. Just as Myspace and Thefacebook were arguably Friendster clones, there were now clones of Thefacebook as well. These copycat sites were opening social networks to target less prestigious schools, the state colleges and even the community colleges that Thefacebook was, up until that point, ignoring. To combat this competition, Zuckerberg accelerated the campus-by-campus rollout so that the clones couldn't steal Thefacebook's thunder. And then there was Myspace itself. The same month that Thefacebook hit 1 million users, Myspace hit 5 million.[35] Zuckerberg was always contemptuous of Myspace, once telling a potential investor that the difference between Myspace and Facebook was the difference between a Los Angeles company and a Silicon Valley company. "We built this to last, and these guys [Myspace] don't have a clue."[36] But then, in July of 2005, Myspace was acquired by News Corp for $580 million. At that point, Thefacebook had only a fraction of the users Myspace did, but if Myspace could command a valuation like that, then Thefacebook was clearly worth some fraction of a very big number.

But the main thing that affected Zuckerberg's thinking was data. From the very first days, Zuckerberg was obsessed with watching how users actually *used* his site. While monitoring the behavior of his users, Zuckerberg was fascinated by the very real info his network could tease out, and how little tweaks he made to Facebook's systems could affect user activity. He had inherited the Google guys' obsession with algorithms. Zuckerberg ran some numbers and realized that, based on things like status updates and wall posts, he could predict with about 33% accuracy whether two members would be "in a relationship" within a week.[37] In theory, he could also predict what movies would be popular, what songs would soon be hits, all from simple posting frequency. That was all pretty cool. But the numbers that really impressed him were those related to user engagement. Usage was off the charts. By the fall of 2005, fully 85% of American college students were members of Thefacebook and 60% returned to the site *daily.*[38] Ninety percent logged in at least once a week.[39] What product or service in any industry got used so obsessively? Parsing the server logs, Zuckerberg and the others could see user behavior that they termed "the trance." Users would log on and then click and click and click and click, browsing people's profiles for hours at a time. "Wanting to look people up is kind of a core human desire," Zuckerberg said around this time. "People just want to know stuff about other people."[40] It was beginning to dawn on him how powerful harnessing that need-to-know was.

It was dawning on other people as well. Venture capitalists and other potential partners were eager to get a piece of Thefacebook. As early as March of 2005, Viacom offered to buy the site for $75 million, thinking that, with its youth demographic, Thefacebook (not Myspace) might be the MTV of the web generation.[41] In lieu of a Viacom buyout or partnership, Sean Parker helped Thefacebook land a $12.7 million investment from the VC firm Accel Partners, which valued the company at around $100 million. The successful investment round was quite an achievement on Parker's part. Google's first major investment round had only valued it at $75 million.[42] Thefacebook was only fifteen months old, but had gotten one of the richest private valuations in Silicon Valley history.

People began to speak in hushed tones about Thefacebook possibly being the "next Google." Zuckerberg himself began playing up this comparison explicitly, recruiting Stanford computer science students behind a homemade sign that read WHY WORK AT GOOGLE? COME TO THEFACEBOOK.[43] Thefacebook's sudden high profile in Silicon Valley, along with its Accel connections, allowed the company to start hiring superstar talent. Steven Chen was such a superstar that he only worked at Facebook for a few months before going on to found YouTube. Facebook stopped renting out "casas de Facebook" and graduated to real office space in Palo Alto.

A final, important sign of the Zuckerberg pivot to taking

Thefacebook seriously came when Sean Parker ceased day-to-day involvement in the company. As *The Social Network* movie suggests, there was, indeed, some sort of incident involving a party Parker hosted, though no charges were ever filed. Thefacebook's new VC investors nonetheless demanded that Parker step down. After a long heart-to-heart between Parker and Zuckerberg, it was agreed that this was actually an opportune moment for change. It was finally time for Zuck to step up and not only take Thefacebook seriously, but take direction of it as well. It was time for him to lead.

This third-time ejection from a startup was more amicable for Parker than the others had been. He got to keep his own sizable chunk of equity. He continued to informally advise Zuckerberg for years afterward. And, crucially, Parker assigned his seat on the company's board of directors to Zuckerberg, giving him control of three seats on the then five-seat board. "That solidified Mark's position as the sort of hereditary king of Facebook," Parker would say. "I refer to Facebook as a family business. Mark and his heirs will control Facebook in perpetuity."[44] Thanks to him.

Oh, and one of Parker's last acts was to secure the domain Facebook.com. Sean had long argued that the "the" in the site's title was superfluous. The company officially became Facebook on September 20, 2005.[45]

■

A LOT OF THE FASCINATION surrounding the story of Mark Zuckerberg has been about the world watching a boy evolve into a legendary entrepreneur and leader. Zuckerberg's trajectory mirrors that of another truly great entrepreneur who dropped out of Harvard to start a company. Bill Gates was almost exactly Mark Zuckerberg's age when he founded Microsoft. He too was called socially awkward and he too had an early reputation for sophomoric behavior that verged on the juvenile. Gates didn't start out to become one of the most successful entrepreneurs of all time. He grew into the role. It was only after he had one of the great business insights of all time—that software was *the* truly valuable nexus point of technology—that he seized his destiny. In truth, the "genius" of Bill Gates was his ability to evolve into the sort of man who could capitalize on his great entrepreneurial insight.

Not being a natural entrepreneur—and then stumbling onto a great entrepreneurial insight—and then having the fortitude, and discipline, and strength of will to become the sort of person who can bring that insight to reality? To me, that's the more fascinating story.

What was Zuckerberg's great insight about Facebook? Well, it was something along the lines of: humans are nothing more or less than highly social primates. Finding out what is happening with your friends and family is a core human desire, right smack in the middle of Maslow's hierarchy of needs. Zuckerberg had once mused that someday somebody was going to make a community site that would satisfy the need to know what's up with your friends—but for the entire planet. And when they did so, they'd be building an amazing company.

Maybe Facebook could be that amazing company.

In short, Zuckerberg began to believe in the power of the product he had already built. And he got strong evidence that he really was on to something thanks to a key new feature Facebook launched.

Over the summer of 2005, the site grew from 3 million members to 5 million.[46] At times, 20,000 new users were joining daily.[47] The site was getting 230 million pageviews daily. Revenues had climbed to $1 million a month.[48] As it had done the previous year, Facebook decided that autumn was the best time to introduce major new features. Before he left, Sean Parker had been advocating for a photos feature to be added to Facebook. Instead of simply a profile photo, Facebook users should be able to share any photo, entire groups of photos, entire photo albums. On Myspace, an ecosystem of third-party companies like Photobucket and Slide had arisen to serve this purpose. And obviously, sites like Flickr showed that people were eager to share photos online. But Parker wanted Facebook to own the experience itself. "The theory behind photos was that it was an application that would work better on top of Facebook than as a free-standing application," Parker says.[49] And maybe if Facebook leveraged what it was already good at—its network effects—it could create something even more powerful still.

Facebook Photos was launched in October of 2005. It was actually a bare-bones application, lacking a lot of the features of more robust apps like Flickr. But it had one key innovation: if you uploaded a photo with a friend in it, you could "tag" them and they would receive a notification that you had posted a photo of them online. Facebook Photos took off right away. Within three weeks, Facebook hosted more photos than Flickr.[50] After a month, 85% of the service's users had been tagged in at least one photo.[51] Zuckerberg and the rest of the team were amazed that an arguably inferior product could so quickly unseat the incumbents. The secret sauce had to be the network effects. Matt Cohler was one of the new wave of hires brought in to Facebook after the Accel investment. "Watching the growth of tagging was the first 'aha' for us about how the social graph could be used as a distribution system," Cohler says. "The mechanism of distribution was the relationships between people."[52]

Again, Facebook didn't invent tagging. It was one of those big ideas floating around the Web 2.0 zeitgeist. But combining tagging of photos with Facebook's unique network of real social connections proved impossibly potent. We're monkeys that like to talk to each other—that like to see and be seen. When someone tagged you in a photo, how could you help but look? Again, the primary way Zuckerberg measured the success of Facebook was by monitoring how often users returned, and how much they clicked on when they did so. After photos, he saw that Facebook's return traffic ramped up in a major way.

"Watching what happened with photos was a key part of what led Mark's vision to crystallize," Sean Parker says. "He was formulating a broader and broader theory about what Facebook really was."[53]

The theory was something like this: human society is all about that small group of people you know and care about. Facebook had succeeded in capturing that, harnessing that, replicating that (at least, for college students). If Facebook really had tapped into one of the most powerful human impulses among college kids, why couldn't it appeal to everyone? A product like Microsoft Windows was used by almost everyone who owned a computer. Billions of users. But a product like Coca-Cola was known to almost every human being alive, was *used* by almost every person alive. Could Facebook and the social graph be that powerful?

▪

IT WAS OVER THE COURSE of the next year, 2006, that Mark Zuckerberg and his company both began to mature. Hiring ramped up. After the blockbuster success of Photos, the company blew through all the storage capacity that had been allotted for the coming six months—in six weeks. Once again, Facebook needed more machines, servers, storage. Facebook raised another capital round to fund this expansion, this time at a $500 million valuation.[54] And in the midst of what was now a full-blown movement around social networks and Web 2.0 generally, an even greater frenzy of interest arose around Facebook. Everyone wanted a piece of the site. And most of the circling sharks wanted to swallow Facebook whole.

Viacom expressed interest in purchasing Facebook again. As did Rupert Murdoch's News Corp. As did Time Warner. For a period of months, it seemed like Zuckerberg took meetings with nearly everyone in the Fortune 100. To outsiders—and also to a lot of people inside the company—it looked like Zuckerberg was planning on cashing in while social networking was hot. Maybe he could flip Facebook for a cool billion or two. Not bad for a few years' work. He could go back to Harvard or retire to the French Riviera. But in retrospect, it seems that Zuckerberg was actually using all this face time with some of the world's most powerful CEOs in order to get a crash-course M.B.A. degree. By fielding offers and partnerships, he could learn the ins and outs of real-world business and finance at the highest levels. When a Viacom executive offered Zuckerberg the use of a corporate jet to fly home and visit his family, it was likely a ploy to get Mark alone for five or six hours so that he could be convinced to sell out. Instead, Zuckerberg spent the entire flight picking the executive's brain about the day-to-day realities of running an advertising-based media company like Viacom.

Facebook still wasn't profitable at this point, so it made sense to a lot of people that Zuckerberg would eventually sell. But there were intriguing signs that there could be a very powerful advertising-based business built off the social graph. One new feature that had been added to Facebook was the ability for businesses or brands to sponsor individual groups and eventually individual pages that would serve as Facebook profiles that users could "friend." Since 2004, Apple had sponsored a popular group that, early on, was Facebook's single biggest revenue generator. When Procter & Gamble sponsored a group for its Crest Whitestrips teeth-whitening product, 20,000 people joined.

From the beginning of the web, all the way through the launch of Google AdWords, the Internet had been monetized on the premise of taking the guesswork out of advertising. Well, on Facebook people were using their real names. They were volunteering their likes and dislikes. You could actually get people to tell you if they were interested in your product or not. It was advertising's holy grail.

In his meetings with Viacom, Zuckerberg mentioned that he believed Facebook was worth $2 billion. Viacom eventually offered $1.5 billion in cash and stock, but only with earn-out and performance conditions, so Zuckerberg declined.[55]

That July, Yahoo offered $1 billion, all in cash. Both Accel and Peter Thiel thought the offer should be seriously considered. But when a board meeting was called to weigh options, Zuckerberg was brief.

"We're obviously not going to sell here," he told the group.

Peter Thiel urged him to at least think about it, pointing out that a billion dollars was a lot of money and there was a lot that he could do with that kind of money.

"I *don't* know what I could do with the money," Zuckerberg responded. "I'd just start another social networking site. I kind of like the one I already have."[56]

The deal was rejected.

Each time an interested acquiring party would enter the picture, Zuckerberg would take meeting after meeting—but he never said yes to a sale. Some of the VCs who had backed Facebook were especially eager for a quick exit, and they began to pressure him intensely. But Zuck could never be persuaded. And if Zuck didn't want to sell, then there would be no sale. Sean Parker had made sure of that. Parker was, in fact, still advising him to stay true to his vision. So was Marc Andreessen. The Netscape founder was just then beginning his new career as a prominent investor in Internet startups. He became a trusted Zuckerberg confidant and eventually joined Facebook's board of directors.

It's possible that Mark Zuckerberg could have sold Facebook during this period, and many people felt he would have been wise to. Friendster hadn't sold at the height of its popularity, and look what had happened to it. Heck, in the dot-com days, TheGlobe had been a "community" site like Facebook. It had IPOed and then ridden the bubble down to pennies on the dollar. Zuck was not unaware of recent history. It's possible a dollar figure could have been floated that he wouldn't have been able to turn down. People began to whisper that Zuckerberg had gotten full of himself, that he was holding out for an impossible valuation, looking to make the deal of the century.

But the truth was, Zuckerberg couldn't shake the feeling that somehow Facebook could be something bigger than a quick flip for a couple billion dollars. "When people say I'm greedy, they're missing that I could already have more money than I'd know what to do with," Zuckerberg told a *Rolling Stone* reporter during these months.[57] He told people he was building Facebook for the long term. He still was nursing the crazy idea that Facebook could become a brand as ubiquitous as Coca-Cola. A billion dollars wasn't cool. What would be cool? A billion *users*. "I don't want to sell," he told one of the more persistent executives looking to buy his company.[58] "And anyway, I don't think I'm ever going to have an idea this good again."[59]

Facebook began to expand overseas, still following the tried and true school-by-school method. In almost every country it entered, Facebook encountered homegrown copycats. In most cases, Facebook quickly trounced the competition. The first steps were taken to expand beyond college users by opening the service up to high schoolers. Since high schools generally don't have school-assigned email addresses, younger users were allowed in only if they were invited by someone they knew who was already in college. This proved irresistible to younger users, and though some existing members grumbled about the "kids" flooding in, the expansion was generally judged to be successful. The next logical step was to expand in the other direction. Already, 60% of members continued using Facebook after graduating and entering the workforce.[60] So, plans were put in place to expand to older users by creating mini networks centered around employers and companies.

At the same time, work began on what would prove to be the single most important feature Facebook would ever develop. When studying the "Facebook trance," the one that led users to click, click, click, Zuckerberg and the others saw that the reason people got so sucked in to the site was that they had to surf around to find out what had changed on every friend's profile page. Users seemed to be most interested in learning what was new. Heck, every time a user simply changed their profile picture, Facebook's engineers could see in the logs that that led to an average of twenty-five new pageviews.[61] If Facebook's key value proposition was the ability to find out what was up with your loved ones, then maybe they could design a better delivery system for this information. This would become the News Feed.

Again, the News Feed built on ideas that were already out there. Every user's profile page would function as a glorified RSS feed, and the News Feed would collect all the updates, photos and status changes that your friends made in one central place—just like a feed reader collected blog posts. You wouldn't have to visit profile page after profile page individually, you could just log in and Facebook would tell you what was new. It would all spool out in one, long, reverse-chronological stream, just like a blog.

But engineering the News Feed was a big ask, from both a design and an architecture perspective. Now when you logged on to the site, Facebook wouldn't just need to call up information from one profile at a time; it would have to pull data from all your friends at once. On top of this, the developers wanted a complicated Google-style algorithm that would sort the updates in the feed based on what it thought you would be most interested in. You'd see updates from people Facebook had noticed you interacted with most often, for example. This was a huge technical challenge—a break from the computational simplicity that, up until that point, Facebook had relied on to avoid Friendster-style slowdowns. So, of course, the News Feed required yet more servers, more databases, more computing power. Facebook would need to ramp up to Google levels of computational sophistication.

In retrospect, the News Feed is so obviously Facebook's "killer application," that it's almost surprising social networks got as popular as they did before the News Feed was even invented. And so, it came as a shock to everyone at Facebook that users *hated* the News Feed. The feature was launched in the early morning of Tuesday, September 5, 2006.[62] By breakfast time, Facebook staffers were deluged with messages of pure outrage. Only one in a hundred postings about the News Feed was positive.[63] Ben Parr, a junior at Northwestern University, created a Facebook group called Students Against Facebook News Feed. It had 700,000 members by that Friday.[64] By some estimates, fully 10% of Facebook users were actively protesting the changes. Most of the complaints about the News Feed centered around the perceived breach of privacy. "Very few of us want everyone automatically knowing what we update," wrote one angry missive, "news feed is just too creepy, too stalker-eque [*sic*], and a feature that has to go."[65]

This was the closest thing to an existential crisis Facebook had ever faced. The history of social networking had shown that users were fickle; they would flock to whatever service best suited their needs at the moment. If you pissed off your users, they would leave you. The reasons Friendster had been abandoned were largely technical, but sites could be brought low by basic design changes as well. A few years after the News Feed brouhaha, Digg would redesign its site and change its voting algorithms in a way that so angered users that they fled, en masse, to a Digg competitor named Reddit. To this day, Reddit is known as the "front page of the Internet," the birthplace of memes and viral culture, while Digg, though still around, is nowhere near as relevant or well trafficked.

So, with the News Feed backlash, panic set in at Facebook HQ. High-level meetings were held among the Facebook brain trust over whether or not to backtrack and shut off the News Feed. Zuckerberg himself quickly penned a note to users, "Calm down. Breathe. We hear you." Privacy controls were hastily coded up to give users better control over what showed up on the Feed and what didn't. But the News Feed was never shut down, even temporarily, because, again, Zuckerberg was watching user behavior and, despite the ruckus, he could see that people were actually using the News Feed as he had intended. In August, before the News Feed, Facebook users viewed 12 billion pages. In October, après News Feed, pageviews were 22 billion.[66] People might claim to hate the feature, but Zuckerberg could see they couldn't stop using it. In fact, the proliferation of anti–News Feed protests was tangible proof that the new feature was working as designed. The whole point of the News Feed had been to surface things happening you might want to know about, he told *Fortune* reporter David Kirkpatrick at the time. "One thing it surfaced was the existence of these anti-feed groups."[67] The News Feed itself had enabled its own backlash to spring up.

The anger blew over eventually, and the News Feed went on to become *the* core feature of Facebook. But it still caused a very real crisis in confidence at a crucial and uncertain time. "If [News Feed] didn't work," Chris Cox says, "it confounded [Zuckerberg's] whole theory about why people were interested in Facebook. If News Feed wasn't right, he felt we shouldn't even be doing [Facebook itself]."[68]

It didn't help that the News Feed near-fiasco came on the heels of a less publicized but no less demoralizing failure from earlier in the summer. When Facebook's work networks were launched, they barely got any attention. Only on army bases, and among U.S. military users, had the workplace networks taken off. But then, military folk were generally the same college-age cohort that Facebook had always been successful with. Adults didn't seem to be interested in the service at all.

The News Feed experience shook Zuckerberg's core faith in what Facebook was all about. And after the failure of work networks, a bigger question hung heavy in the air: was Facebook really just for kids after all? If so, then Zuckerberg's great insight, that his social graph was a useful thing for everyone on the planet, was mistaken. "It was the most wrong he'd ever been at Facebook, and the first time he'd ever been wrong in a big way," early Facebook executive Matt Cohler said of this period of doubt.[69] If Zuckerberg was wrong about these things, had he also fundamentally misjudged the big, world-changing value of Facebook to begin with? In that case, maybe a $1 billion sale wasn't such a bad outcome after all. They had already conquered the high school market. Myspace still had a lead in the overall twenty-something demographic. If older users couldn't be enticed to join, there wasn't any more low-hanging fruit to be had in terms of harvesting growth.

And it was at this exact moment, in September 2006, that Yahoo came back and renewed its $1 billion all-cash offer. Yahoo's lawyers did due diligence on Facebook's finances and operations, and an acquisition was agreed to in principle. Given the stumbles of the past few months, nearly everyone was now in favor of a sale—especially the VC investors, but plenty of rank-and-file Facebook employees as well.

Everyone, that is, except for Mark Zuckerberg. And even he was beginning to waffle.

"We almost took the offer," Sean Parker would later say.[70] It was seemingly the only time the pressure to sell got to be too much for even Zuckerberg to resist.

But before agreeing to sell, Zuckerberg wanted to take one last crack at opening Facebook up to everyone. Once more, he played for time, dragging his feet on the acquisition talks, taking meeting after meeting but not actually pulling the trigger on the Yahoo deal. He wanted to see if his gut instincts about Facebook were right.

Perhaps—perhaps the work groups had failed because they were the wrong paradigm. Maybe Facebook had used its tried-and-true network-by-network expansion trick one time too many. Maybe explicit networks were less important outside of a school setting. The people who graduated college but still continued to use Facebook just took the network with them, even when they moved away from campus. Perhaps the thing to do was just throw registration wide open and let everyone in. That way users could grow their networks organically.

The engineers borrowed an idea from Sean Parker. Plaxo had grown by searching users' existing address books and email programs to invite people to join and make connections. An "Address Book Importer" was designed to go into your Hotmail or Gmail account and search for other users who were already on Facebook. That way, new users would be greeted with a slew of people they already knew when they signed up and needed to begin populating their network. The importer would serve up friend connections on a platter, and anyone not on the service could be invited to join via the same mechanism.

It was one last roll of the dice. One last gamble, where failure still meant $1 billion and success meant—well, who knew?

■

OPEN REGISTRATION WAS LAUNCHED on September 26, 2006, mere weeks after the News Feed debacle. Prior to open registration, new users were joining at a rate of about 20,000 a day. A few weeks after opening up Facebook to everyone, that number had changed to 50,000 a day, and rising.[71] Growth in Facebook's user numbers began to look like a hockey stick going only steeply upward. Over the next year, Facebook would rocket past 25 million registered users, and around 6 million of those would be older-than-college-age users; 200,000 of those would even be people over age sixty-five.[72] If you were a postcollege adult during this period, you might remember this moment. One day, Facebook was just a thing you had heard of. The next day, everyone you knew was on it. Some day after that, your mother and even your grandmother were members.

The one personal anecdote I'll share in this book: that summer of 2006 was my ten-year high school reunion. It was an important event. Many of my classmates had lost touch with each other. There were a lot of "Wow! What happened to you?" conversations. And then, just a few months after we got together, open registration happened, and we all found each other again on Facebook. Soon we were all even connected with classmates who hadn't been able to make the reunion.

Ten years later, our twenty-year high school reunion in 2016 was less of an event. It was more of "Hey, I saw the photo of your new car yesterday" than it was "Where have you been?" After all, thanks to Facebook, I now get updates about everyone on an hourly basis. I know that my senior-year chemistry lab partner just got back from a trip to China and that the oldest child of the girl I kissed in sophomore year just broke his arm skateboarding. There is a very clear demarcation point to my social life between pre-Facebook times and post-Facebook times, and it felt like the change happened overnight.

■

IT SORT OF DID HAPPEN OVERNIGHT. From its launch in 2004 until open registration in 2006, Facebook grew to around 8 million users.[73] One year after open registration, Facebook had 50 million active users.[74] By the end of 2008, there were 145 million people on the service, 70% of them outside the United States.[75] The next year, there were 350 million users in 180 countries. After open registration, the social-networking wars were over. Myspace, and every other social network, would become distant memories.

It turned out that Mark Zuckerberg was right. Connecting everyone together—almost the original premise of the web itself—was an incredibly useful and valuable thing indeed. Zuckerberg is the twenty-three-year-old who turned down a billion dollars because he thought he was sitting on an idea that was even bigger. The gamble has paid off (at the time of this writing) to the tune of a nearly half a trillion dollars in market value. It helped that advertising against everyone's personal lives also proved to be lucrative, and that the reverse-chronological scrolling mechanism of the News Feed proved to be perfectly suited for the coming age of mobile computing. But none of that would have been possible had Zuckerberg not matured into the sort of businessman who could make such a gamble. The fact that he did is the entrepreneurial story of our age.

16

THE RISE OF MOBILE

Palm, BlackBerry and Smartphones

In the technology world, the ultimate success of a new idea is very much dependent on timing. Even great ideas that are quite obviously "the next big thing" can fail to deliver on their promise because the underlying technology or infrastructure isn't mature enough yet. Streaming video was supposed to be big, going back to the days of Real|Audio and Broadcast.com, but it took the example of Napster and the advent of broadband Internet connections before YouTube could take off. SixDegrees couldn't succeed because it was birthed in a world before ubiquitous digital cameras. Facebook got the timing right on that detail, but it also cracked the social-networking code by achieving critical mass just as another key technology was having *its* breakthrough moment.

For many long years, mobile computing was an idea before its time. Dozens of attempts to jump-start mobile computing as an industry crashed and burned without gaining widespread adoption. After the PC revolution and just prior to the dot-com era, there was a brief fad in Silicon Valley for handheld computers. It was the logical next step: once there was a computer on every desk, why not put one in every pocket? In the late 1980s and early 1990s, there was a minibubble as investors rushed to fund dozens of handheld, mostly pen-based, computer startups, both on the software and the hardware side. GO Corp. burned through $75 million dollars in VC money in an attempt to become the Microsoft of handhelds by creating the operating system standard for pen computing. A company called GeoWorks attempted something similar with its GEOS. General Magic was an Apple spin-out that Pierre Omidyar of eBay; Tony Fadell, the father of the iPod; and Andy Rubin, the inventor of the Android operating system, all worked at before going on to fame and fortune elsewhere.

Before it was spun out, General Magic was one of two top-secret research-and-development teams[1] inside Apple Computer. The other team, which remained in-house, was called Newton, and it would be responsible for the highest-profile early handheld computing device. As the 1980s turned into the '90s, the Newton team was working on a tablet computer the size of an eight-and-half-by-eleven-inch sheet of paper. This experimental device weighed about eight pounds but was only three-quarters of an inch thick. Named Figaro, the machine was navigated using a stylus on a grayscale screen, had three processors, an internal hard drive and wireless networking and got about ten hours of battery life. Oh, and it cost about $8,000 to produce.[2] Per device.

In early 1991, a young Apple marketing executive named Michael Tchao convinced Apple's then-CEO, John Sculley, to switch gears and have the Newton team work on a smaller, ultraportable computer—one that could fit comfortably in a person's palm. Sculley became an evangelist for the idea of a near-pocketable computer, a category of devices he termed personal digital assistants, or PDAs. At the 1992 Consumer Electronics Show, Sculley declared that there would soon be a "$3.5 trillion" market for such devices.[3]

The result of this strategic pivot, the Newton MessagePad, was released to the public on August 2, 1993. It cost $699, ran on four AAA batteries and weighed 0.9 pounds. But at 7.24 inches by 4.50 inches (about the size of a VHS cassette) it was hardly pocketable, except in the most generously sized pockets.[4] With optional add-ons, you could send faxes and (eventually) email using a wired modem. But the main features of the Newton were its productivity apps, including a calendar, address book, to-do list and notepad. It had no keyboard, instead boasting a touchscreen that you interacted with using the included stylus. The intention was, you would write on the Newton just as you would if you were writing on a piece of paper. The software would interpret your handwriting and turn it into on-screen text.

Or, at least, it was supposed to. The Newton was ultimately done in by its notoriously flaky software, which, more often than people could tolerate, simply refused to recognize what had been written. Oddly enough, the longer the word, the better the software was at translating, because longer words gave the handwriting interpreter more information to work with. The Newton struggled primarily with shorter, monosyllabic works like "or" and "the."[5]

The Newton's software was supposed to learn your handwriting over time, but *PC Week* complained that, "The Newton is almost worthless . . . basically shelfware. After three weeks, it still couldn't consistently differentiate my 1's from my t's."[6] Other reviews were just as scathing. "Apple promised too much and failed to deliver a useful device," wrote the *New York Times*.[7] In a classic example of a rolling PR catastrophe, after the Newton came out, the comic strip *Doonesbury* spent a week turning the Newton's handwriting recognition foibles into a national joke.

Apple had expected to sell 1 million Newtons in the first year. It sold only 85,000.[8] Subsequent models would improve immeasurably, especially the second-generation device that was Jony Ive's first assignment after being hired at Apple. But it was too late. In the court of public opinion, the Newton could never overcome its poor reputation.

■

THE NEWTON'S HIGH-PROFILE failure took the entire nascent handheld computing market down with it. Most of the handheld startups went out of business in the coming years, just as Silicon Valley was turning its attention to the web. One of the handhelds that made it to market, only to be dragged down in the Newton's wake, was the Zoomer, a $700, one-pound pen computer that debuted in October of 1993, selling only 60,000 units before being discontinued.[9] But even a firsthand brush with failure could not kill the dream of a "computer you could carry in your pocket," not for Jeff Hawkins, the inventor of the Zoomer. Hawkins had founded Palm Computing in January 1992 to produce the Zoomer, and even after its first product failed in the marketplace, Hawkins and a small band of Palm loyalists merely went back to the drawing board and began sketching out a follow-up device.

Hawkins had a hunch that handhelds had attempted to do too much, had been too complex, too ambitious. He intuited that people didn't necessarily want a second computer, they wanted an *accessory* to their existing computer. So, he focused on only a few key use cases for his new device: a calendar, an address book and a memo pad. These applications would be designed to sync to regular computers when the device was connected by wire; when out in the "wild," as it were, the device would stick to its primary, simple task: helping the user stay organized.

Hawkins began carrying a rectangular piece of balsa wood, about the size of a deck of cards, around Palm Computing's offices. With this dummy mock-up, Hawkins tested out the ideal dimensions that would allow a handheld device to be useful in everyday situations. The resulting product would be known as a PalmPilot (though it had various names due to branding and trademark issues over the years). By sticking to Hawkins's ethos of simplicity, not only was the Pilot eminently pocketable (it was about a third the size of the Newton and weighed 5.5 ounces); it could also hew to a $300 price point, thereby making it seem like a logical desktop or laptop accessory.

Palm would sell 1 million Pilot units in eighteen months on the market, thereby becoming the fastest-selling computing device in history.[10] It was still a pen-based gadget—there was no physical keyboard—but Hawkins had solved the handwriting input issue that had beguiled the Newton by creating a single-stroke shorthand alphabet known as "graffiti." This improvised input language worked well and the Pilot proved useful, especially to businesspeople on the go, with a simple plug-in-and-sync interface, much akin to what would later become commonplace with the iPod and iTunes system. By 2001, Palm had sold 21 million of these pocket computers and secured a 70% market share of a reborn PDA market.[11]

In Canada, another small company took notice of the rebirth of the pocket computer and decided to come at the market from a different angle. If Jeff Hawkins focused on the simplicity of productivity and organization while on the go, Mike Lazaridis, the founder of Research In Motion (RIM), focused on communicating while on the go. In 1996, RIM launched the Inter@ctive Pager, a two-way wireless messaging device. Initially, it was just a glorified pager. But Lazaridis and the RIM engineers concocted clever ways to hook into personal and corporate email systems and eventually, RIM was delivering, essentially, email in your pocket. The first Inter@ctive Pager, the 900, and its successor, the 950, released in September of 1998, shared the Palm ethos of simplicity, pocketability and utility on the go. Measuring 2.5 × 3.5 inches and weighing 4.5 ounces, the RIM pagers mimicked the PalmPilots in their form.[12] "Everyone else was trying to add a radio to a PDA," recalled Dr. Peter Edmonson, RIM's chief radio engineer. "Whereas Mike's mindset was how to add a PDA to a radio."[13]

RIM's devices were designed to be "always online" as opposed to syncing to a computer occasionally, as the PalmPilots were designed to do. Email was "pushed" to RIM's gadgets over the wireless network, so you didn't have to plug in to find your messages; your messages found you, wherever you happened to be. When you got something new in your inbox, the device would buzz and a red LED would indicate that you had a new message to read. Because RIM had previous experience working with radios and wireless networks, its pagers were fast and incredibly energy-efficient. The 950 could last for three weeks on a single AA battery. And RIM didn't mess with the touchscreen technology that Palm was so married to. Instead, RIM innovated tiny, fully functional keyboards designed to be used with one's thumbs. "For me, it was all about keyboards," Lazaridis has said. "Jeff [Hawkins] went off and did touch screens. I went off and tried to develop something with a keyboard."[14]

On January 19, 1999, RIM launched the 850, the first device that would carry the name BlackBerry.[15] It was also the first mobile device that synced completely with email systems, so sending and receiving an email on the go was as seamless as communicating from your computer. If you sent an email from your BlackBerry, it showed up in your Sent folder when you got back to your desk. Messages read on the BlackBerry were marked as read on your computer, and vice versa. RIM also began integrating more PDA-like functionality into the BlackBerry and subsequent models, so that eventually they had all the functionality of a PalmPilot, but with comprehensive messaging capabilities.

The BlackBerry's marketing tagline was "Always on. Always connected." As the 1990s turned into the 2000s, among a class of professionals for whom never being "out of the loop" was of paramount importance, the BlackBerry took off like wildfire. "It very quickly became a status symbol," recalled wireless research consultant Andy Seybold.[16] BlackBerry proved popular on Wall Street, among lawyers, in Hollywood. When you watch old video of the AOL/Time Warner merger announcement, you can see Jerry Levin and Steve Case checking their BlackBerrys to see how the news was affecting the stock price of their respective companies. In the disputed election of 2000, the Gore campaign managed its response to the "hanging chad" situation, minute-by-minute, over their BlackBerrys. During the terrorist attacks of 9/11, most cell service went down, but BlackBerry users could still get their messages out. Congress subsequently bought BlackBerrys for every senator, representative and thousands of Capital Hill staffers, such was BlackBerry's reputation for keeping people in the know.[17] When Oprah Winfrey broadcast one of her annual "Oprah's Favorite Things" specials, she gushed: "I cannot live without this. It's with me everywhere I go. It's called a BlackBerry. It's literally changed my life."[18]

What so entranced these early adopters of the BlackBerry was just that ability to always be connected to information. It's the reality we're all familiar with today: the phenomenon of never being out of touch. But in the early 2000s, this was a new experience. For BlackBerry users, there was never a moment when they couldn't be reached, when their device didn't beckon to them with a new alert of someone trying to reach them or some new piece of information to digest *right away*. BlackBerry users were the first people to confront the social etiquette implications of conversations and person-to-person interactions being interrupted by digital notifications. And they were the first to wrestle with the uniquely obsessive mindset that an always-on information device can engender. This pull of the "now" only got worse as BlackBerrys eventually gained web-browsing functionality and new applications such as the BlackBerry Messenger instant messaging service. The devices earned the sobriquet "CrackBerry" because users seemingly couldn't tear themselves away.

"It should be reported to the DEA," Intel chairman Andy Grove told *USA Today*.

"It is the heroin of mobile computing," Marc Benioff, CEO of Salesforce, said in the same article. "I am serious. I had to stop. I'm now in BA: BlackBerry Anonymous."[19] Communication, as it so often did over the course of the Internet Era, proved to be the killer application for mobile computing. But then, heroin is a "killer" application as well.

Palm eventually released handsets with radios that mimicked the BlackBerry and enabled messaging, especially the popular Palm VII in 1999, which added email to Palm's traditional organizer applications. But by 2005, RIM had replaced Palm as the largest seller of pocketable computers.[20] And at that point, the handheld computing market that both Palm and RIM were chasing was careening headlong toward something even greater than any of the mobile computing pioneers could ever have imagined.

■

PDAS, PAGERS, EVEN MP3 PLAYERS, were the hot consumer electronics products in the early 2000s. But in this burgeoning world of electronic devices that were competing for room in your pocket, there was only one undisputed king: the cell phone. Other devices might be able to capture the imagination of certain market segments, but cell phones were seemingly for everyone. There were 100 million cell phone users worldwide as early as 1995. By 2001, that number surpassed 1 billion. And midway through the decade, nearly a billion handsets were being sold every single year.[21] Because phones were clearly the most popular pocket devices on the planet, it made sense that the features that were turning handheld gadgets into must-have objects of envy began to be subsumed into phones as well.

The very first smartphone was the Simon Personal Communicator, which was developed by IBM back in 1992. On sale to consumers for just six months, from 1994 to 1995, retailing for $895, the Simon had almost all the components that we would recognize in a modern smartphone. It could send and receive cellular calls, of course, and it could also send and receive pages wirelessly. It could do email and fax, but those required the user to dial in via a landline. It could sync via an adapter cable to a computer, and could therefore store and work with data files. The majority of the device consisted of a touchscreen, where a row of icons could be found that summoned up an array of apps, including an address book, a calendar, an appointment scheduler, a calculator, a world clock and an electronic notepad. The Simon weighed slightly more than a pound, but at 8 inches long by 2.5 inches wide by 1.5 inches thick, it was more of a brick than a pocketable device. It had a rechargeable battery, and even an expansion slot for adding more memory. The plan was to eventually add additional hardware and software features like maps, a GPS module, real-time stock quotes and more.

Unfortunately, the Simon never got there. IBM sold only 50,000 Simons before discontinuing the product. "It's all about time frames," says Frank Canova Jr., who led the device's development at IBM.[22] "The Simon was ahead of its time in so many different ways."[23] All the features of a modern, communicating, mobile computer were already there, but the world just wasn't ready for it.

Too soon.

But as Palm and RIM found success with PDAs and pagers, the broader cell phone industry had second thoughts about pocket computing. The 800-pound gorilla of the cell phone industry in the late 1990s and early 2000s was Finland's Nokia. In 1996, it released the 9000, the first of its Communicator series of phones. The Nokia

9000 opened up, clamshell-style, to reveal a full QWERTY keyboard. It had a web browser as well as digital camera connectivity. It could make calls, of course, and send messages, and had the now-usual suite of contacts, notes, calendar and calculator apps. But since cellular data plans were rare and expensive, the Communicator series was not a mainstream success.

Too soon.

The first cell phone to be explicitly called a "smartphone" was the Ericsson R380, released in 2000. Its lid flipped open to reveal a full touchscreen for web browsing, email, apps and games. Other manufacturers soon followed Nokia and Ericsson's lead, releasing a wide range of devices, all of which married PDA and messaging function to phones, some going the Palm route, with touchscreens, and some the BlackBerry route, with thumb-friendly keyboards. And the handheld pioneers themselves also joined the fray, with the Palm Treo line of smartphones beginning in 2002, and the BlackBerry Quark that Oprah called one of her favorite things targeted toward mainstream consumers beginning in 2003.

Then, a whole slew of manufacturers jumped into the smartphone game. In order to stand out from the crowd, every imaginable feature started getting crammed into phone handsets. The first phone with integrated GPS was released in 1999. Japanese consumers were buying phones with integrated digital cameras as early as 2000. Many phones began to offer rudimentary web browsers and even streaming video by the middle of the decade.

Too soon.

All through the first half of the 2000s, mainstream consumers collectively yawned at the explosion of smartphone and mobile computing features. By 2005, there were only 3.5 million smartphone subscribers in the United States.[24] As late as 2006, only around 6% of the 150 million phones shipped in North America were "smart."[25] Even though it was used by 85% of Fortune 500 companies, RIM didn't reach a million subscribers until 2004.[26] Palm's sales actually began declining, beginning in 2000.

The entire computer, electronics and technology industry was converging on one singular device, one transcendent product that would seemingly be everything to everybody. And yet, few people seemed to care. All of these new features, all of these new technologies and computing innovations were converging inside the cell phone, pointing to a world of always-on, always-connected, always-updating information, but aside from those CrackBerry addicts and hard-charging professionals, most people didn't see the point.

Back in 1998, Steve Jobs famously told a *Businessweek* reporter that "a lot of times, people don't know what they want until you show it to them."[27] In the case of the smartphone, in the case of the technology that would soon bend the entire arc of modern life toward the ubiquity of mobile computing, that would certainly prove to be true.

17

ONE MORE THING

The iPhone

The logic that was driving device manufacturers to cram a kitchen sink's worth of technology into the singular form factor of the smartphone was simple. Why carry multiple devices around when you could just carry one? Why would you want a PDA *and* a messenger? You wouldn't. So, Palms gained messaging capabilities. Why would you want to carry a messenger *and* a cell phone? So, BlackBerrys gained the ability to make phone calls.

But what about that other device that, for about half a decade, was also taking up space in everyone's pockets? Clearly, if you could store music on your cell phone, you wouldn't need to carry around an MP3 player. And nobody was more aware of the cold logic of this than Apple.

"The iPod was selling. It was selling better and better. It was probably 50% of our sales [by the mid 2000s]," says Scott Forstall, a senior Apple executive at the time. "And so, we kept asking ourselves, 'What concerns do we have about the iPod's success, long-term? What will cannibalize iPod sales?' And, one of the biggest concerns was cell phones."[1]

Apple had a vested interest in preventing cell phones from eating its iPod/iTunes lunch. And, as the example of the iPod Nano supplanting the iPod Mini illustrated, Apple could be ruthless when it came to killing its darlings. At the same time, it can't be underestimated how much the success of the iPod changed Apple, altering not only what the company thought of itself, but also changing the very notion of the type of business it could be.

Another Apple executive, Phil Schiller, said that the iPod completely changed Apple's opinion about its own raison d'être. "People started asking, 'Well, if you can have a big hit with the iPod, what else can you do?' And people were suggesting every idea—make a camera, make a car—crazy stuff."[2]

So, an iCamera? Maybe an iTelevision? Like MP3 players, these were standard, stand-alone consumer electronics products. You didn't need anyone's permission to sell devices like these to the masses. Apple could probably gin up the best damned camera anyone had made since George Eastman and potentially blow a whole new industry completely out of the water.

But a cell phone was an entirely different proposition, because it required working with the carriers that controlled the cellular networks in order to make a phone. The carriers decided which devices would be allowed on their networks. They decided the technology those devices could use. They even decided what type of features those devices could have. In short, the cell carriers dictated to the device manufacturers, with the end result being that, in spite of the explosion of features brought on by the smartphone revolution, innovation in the cell-phone space was actually incremental and bureaucratic.

Apple was not a company that liked bureaucracy. Furthermore, Steve Jobs had only recently dragged the music industry kicking and screaming into the twenty-first century. He didn't relish the prospect of having to cajole and educate another recalcitrant group of backward-thinking companies. At the All Things D conference in 2004, the venture capitalist Stewart Alsop Jr. virtually begged Apple to make a phone. Jobs demurred. "We've visited with the handset manufacturers and we've talked to the Treo guys [Palm]," Jobs said. "They tell us horror stories."[3] At the same conference the following year, Jobs outlined the problem, as he saw it. "The carriers now have gained the upper hand in terms of the power of the relationship with the handset manufacturers," Jobs said. He described how manufacturers would get thick books of product and network specs from the carriers, which essentially dictated everything a cell phone could be, down to the last screw and wire. That wasn't Apple's MO. "The problem with a phone is that we're not very good going through orifices [like the carriers] to get to the end users,"[4] Jobs said.

Still, the momentum of technologies converging into the singular device of the smartphone was hard to miss. And the poor state of the art when it came to cell phones was something of an irresistible challenge to a company that was feeling its oats and eager to solve big problems. "We looked around," said Forstall, "And we noticed that almost everyone around us had phones. And everyone was complaining about their phones. And we thought, 'Could we build something better?' "[5]

Wary of working with the carriers directly, but looking to protect the iPod/iTunes franchise, Apple dipped its toe into the cellular waters by partnering with one of the existing handset makers, Motorola, in early 2004. Apple would merely license the iTunes software, while Motorola would design the hardware, and—most important—deal with the carriers. "We thought that if consumers chose to get a music phone instead of an iPod," remembered Tony Fadell, Apple's executive in charge of the iPod franchise, "at least they would be using iTunes."[6] Apple settled on working with Motorola because it dominated the handset business with its recent release, the RAZR flip phone. The RAZR was a "dumb" phone, not a smartphone, but it was thin, sexy, well designed. In short, it was the sort of product Apple was willing to associate itself with. The RAZR was a huge hit, selling 50 million units in just two years.[7] So, for its first, experimental foray into cell phones, Apple thought it would be injecting its software magic into one of the hottest devices around.

But instead of producing a music-enabled RAZR, Motorola ended up delivering the clunky ROKR. Motorola took eighteen months to deliver this candy bar–style device and it was fatally, almost ridiculously, flawed. It reeked of a handset that was designed by committee, something antithetical to everything Apple stood for. It could hold only one hundred songs, making it the most limited MP3 player Apple had a hand in producing. Within a month of going on sale, customers were returning the ROKR at six times the industry average for a cell phone.[8] *Wired* magazine asked of the ROKR, "You Call This the Phone of the Future?"[9]

"This is not gonna fly," Jobs told the iPod guru Fadell. "I'm sick and tired of dealing with bozo handset guys."[10]

The star-crossed ROKR had been developed in partnership with the wireless carrier Cingular (soon to become AT&T after a series of mergers). At the time, Cingular was struggling to compete with industry leader Verizon. While the ROKR was being developed, Cingular executives began to try to convince Steve Jobs to create an Apple phone exclusively for their network. At first, Jobs refused even to listen to Cingular's entreaties, instead toying with the idea of launching a stand-alone, Apple-branded cellular network. "Jobs hated the idea of a deal with us at first," Cingular executive Jim Ryan said. "Hated it."[11] But Ryan stressed to Jobs the headaches involved in becoming a carrier, not just a hardware maker. Indeed, the customer service, logistical, technical and reliability issues of operating a nationwide cellular network were something Apple had zero experience with. "Funny as it sounds, that was one of our big selling points to [Apple]," Ryan recounted. "Every time the phone drops a call, you blame the carrier. Every time something good happens, you thank Apple."[12]

At the same time Cingular was trying to sell Jobs on the idea of making an Apple phone, a handful of Apple execs, especially Mike Bell and Steve Sakoman, were making the argument for an "iPhone" as well. Bell sent Jobs a long, thoughtful email on November 7, 2004, outlining all his arguments. "I said, 'Steve, I know you don't want to do a phone, but here's why we should do it." Jony Ive had some great iPod designs in the pipeline, Bell reported. All they had to do was pick one, throw some patented Apple software in it, add a cellular radio, and make their own phone. "He calls me back about an hour later and we talk for two hours, and he finally says, 'Okay, I think we should just do it.' "[13]

The deal Apple would cut with Cingular/AT&T would take a

year to finalize, but it alleviated almost all of Jobs's concerns. In exchange for an exclusive right to an Apple phone on its network, AT&T would grant Jobs carte blanche to design the phone as Apple saw fit. It would be completely Apple-branded and AT&T would have no say in the features or services the phone offered. As icing on the cake, Apple would get a share of the monthly cellular data payments users would have to cough up to use the device.

■

BY EARLY 2005, an iPhone was in development. In Cingular, Apple had a partner that would allow it to design a phone as Steve Jobs felt it should be done. But Apple still had zero experience designing a phone, so how the device would turn out in the end was entirely up in the air.

As Mike Bell suggested, the most logical thing to do was to simply add radios to existing iPods. iPods were beloved. Apple was already manufacturing them by the tens of millions. How hard could it be? In a high-level meeting, Jobs signed off on the plan, saying, "We're going to do this iPod-based thing, make that into a phone because that's a much more doable project. More predictable."[14]

The phone project gained the internal code name Purple. Early prototypes were patched together that were merely that: existing iPods, with attached cellular and WiFi radio antennas. But as straightforward as the concept was, iPod+phone simply didn't pan out in real-world use cases. The problem was that the iPod's vaunted click wheel—while a brilliant user interface breakthrough when selecting songs from a list of albums—was not ideal for dialing a phone, much less inputting things like text messages. "We were having a lot of problems using the wheel. . . . It was cumbersome," Fadell told Walter Isaacson in his biography of Jobs.[15]

An Apple engineer named Andy Grignon was tasked with demoing one of the first iPod prototypes to include WiFi. To browse the web on the iPod's tiny screen, "You would click the wheel, you would scroll the web page, and you could click on it, and you could jump in," Grignon said. "And [Jobs] was like, 'This is bullshit.' He called it right away. . . . 'I don't want this. I know it works, I got it, great, thanks, but this is a shitty experience.' "[16]

Fortunately, there was another possible solution waiting in the wings. It just so happened to be an idea that Jobs had also dismissed—at least a first.

Back when Steve Jobs returned to Apple and saved the company from oblivion, he did so, in part, by drastically reducing Apple's focus to only a few core products and technologies. Apple engineers continued to work on skunkworks projects, but they were forced to do so on the down-low, lest Jobs learn of their efforts and shut them down. In the early 2000s, a cadre of Apple engineers was interested in exploring new computer interfaces beyond the typical keyboard or mouse. To stay off Jobs's radar, the engineers often met in Apple's abandoned user-testing lab. In the Steve Jobs era of Apple, focus groups and user testing were superfluous. Only one person (Jobs, of course) decided whether products were worth producing or not.

The secret group was focused on the future of traditional computing, not gadgetry. "Phones weren't even on the table then," says Joshua Strickon, one of the underground engineers. "They weren't even a topic of discussion."[17] The group was more interested in the sort of computer wizardry that had been shown off in the recent sci-fi film *Minority Report*. Gestural input, waving your hands around to manipulate data, etc. The group became fascinated with technology from a small Delaware technology company called FingerWorks. FingerWorks produced a plastic touchpad that allowed users to interact with data directly, in a manual, tactile way, using what was known as multitouch finger tracking.[18]

Someone brought in a Mac, set up a projector over a table and positioned the FingerWorks trackpad beneath it. Soon there was a table-sized demo that showed how a user could interact with a full computer operating system using just their hands. The group shared their demo with Jony Ive and the rest of Apple's industrial design team. Ive was more than impressed.

Dubbing the demo the "Jumbotron" since it was the size of a Ping-Pong table, he told the team to wait until the time was right to show it to Jobs. "Because Steve is so quick to give an opinion, I don't show him stuff in front of other people," Ive explained later. "He might say, 'This is shit,' and snuff the idea. I feel that ideas are very fragile, so you have to be tender when they are in development. I realized that if he pissed on this, it would be so sad, because I knew it was so important."[19]

Indeed, when Ive finally did demo the Jumbotron for Steve, in the summer of 2003, "he was completely underwhelmed," says Ive. "He didn't see that there was any value to the idea. And I felt really stupid because I had perceived it to be a very big thing."[20]

But every so often, ideas that Steve Jobs dismissed at first could grow on him over time. One day's stupid idea could become tomorrow's brilliant breakthrough. "As far as I know," says Brian Huppi, one of the engineers responsible for the Jumbotron, "Jony showed him the demo of multitouch and then it was clicking in his mind. . . . Steve does this, you know: He comes back later and it's *his* idea."[21]

The idea clicking in Steve's mind was the notion that somehow the multitouch technology could be used to solve the phone problem.

"I was sitting with Steve at lunch one time," remembered Scott Forstall. "And Steve said, 'Do you think we could take that demo we're doing with the tablet and multitouch, and shrink it down to something big enough—or small enough—to fit in your pocket?' "[22]

Work on the Jumbotron had continued in fits and starts, with the assumption that the end result might be some kind of tablet, what would eventually become the iPad. But in late 2004, the word came down from Jobs officially: "We're gonna do a phone. There's gonna be no buttons. Just a touchscreen."[23] Apple purchased FingerWorks for the multitouch technology, and soon the phone project was split into two competing tracks. P1 (shortening the Purple designation) became the code name for the existing iPod+Phone version. P2 became this new, multitouch, shrunk-down tablet idea.

In order to make either version work, Apple would need to design the software as well as the hardware. Forstall, who had worked on the Mac's OS X operating system, was put in charge of software development. With Jobs's famous obsession with secrecy, Forstall was told he couldn't hire anyone from outside the company to work on his part of the project; but he was nonetheless free to pick liberally from internal talent. Forstall didn't tell recruits what, exactly, they would be working on. He only divulged that they would be expected to "give up untold nights and weekends and that you will work harder than you have ever worked in your life."[24]

As eventually became standard practice at Apple, the phone team was segregated even from other Apple employees. "The team took one of Apple's Cupertino buildings and locked it down," Forstall would recall in later court testimony. "It started with a single floor with badge readers and cameras. In some cases, even workers on the team would have to show their badges five or six times."[25]

The floor became known as the "purple dorm."

"On the front door of the Purple Dorm we put a sign up that said 'Fight Club' . . . because the first rule of that project was to not talk about it outside those doors," Forstall testified later.

Early on, the software teams came up with the user interface features that would go on to make the eventual iPhone feel so magical. There were the features inherent to multitouch, of course, like pinching or widening your fingers to zoom in and out on pictures or graphics. And scrolling through items was a simple as flicking one's finger up or down the screen. Forstall himself came up with the idea of the double tap to zoom in on text when browsing the web. An Apple UI whiz named Bas Ording came up with the famous rubber band effect, whereby the screen would seem to bounce when a user scrolled to the bottom. To organize the various programs the phone would need, the now-familiar grid of icons was settled upon relatively quickly. Little, squarish, chiclet-like icons seemed to make the most sense for fingers to target. "It's funny, the look of smartphone icons for a decade to come was hashed out in a few hours," says Imran Chaudhri, a senior Apple designer.[26]

Meanwhile, the P1 design was still in the running, pushed by Fadell's iPod team. Given the limitations of the scroll wheel, some were pushing for a hardware keyboard like that on the BlackBerry. "It was definitely discussed," Fadell said later. "It was a heated topic."[27] The software-only keyboard was, in fact, proving to be the biggest problem arguing against the P2 track. It was one thing to implement typing on a multitouch keyboard as big as a table. It was another thing entirely to type on a tiny piece of glass only a few inches in surface area.

Still, after six months of running a bake-off between the P1 and P2 options, Jobs was ready to pick a horse and go with it. "We all know this is the one we want to do, so let's make it work," Jobs said, pointing to the touchscreen P2.[28] It was a risk to go with the untested technology, especially with the keyboard issue still unresolved, but in the end, the possibilities inherent in multitouch were just more exciting.

▪

IF THE SOFTWARE was problematic, the hardware was even more so. It didn't help that the engineers working on the hardware were forbidden from seeing the software that they were ostensibly designing for—and vice versa. The main issue was that Apple simply hadn't dealt with the basic realities of cell-phone design before. Apple also had no experience with the rigorous testing required to (a) function on Cingular's network and (b) pass FCC muster. Handset manufacturers usually left this process to the carriers to sort out, since they were the ones that knew their networks the best. But Apple was keeping AT&T at arm's length, jealously guarding its design even from its nominal partner. And so, the team instigated an intensive "dogfooding" regimen among Apple employees. In technology parlance, dogfooding is when you test your beta product yourself, eating your own dogfood, as it were, in order to work out the bugs. Apple engineers were instructed to live on their iPhones exclusively, to catch bugs in every possible use case.

Dogfooding was coupled with a signal-testing regimen that was nothing if not ad hoc. Often, the process involved little more than driving the phones around in cars and finding dead zones and diagnosing dropped calls on the spot. "Sometimes it would be 'Scott [Forstall] had a call drop. Go figure out what's going on,'" an engineer named Shuvo Chatterjee remembered. "So, we'd drive by his house and try to figure out if there was a dead zone. That happened with Steve too.

There were a couple of times where we drove around their houses enough that we worried that neighbors would call the police."[29]

Parallel to these efforts, the industrial design team under Jony Ive was churning out prototype after prototype. One intermediate hardware design that Ive was particularly fond of was based on an iPod-like design from the P1 track. The device was made of brushed "aluminium," of course, so Jobs and Ive loved it. But in this instance, the master aesthete had to bow to the laws of physics. "I and Ruben Caballero [an antenna expert] had to go up to the boardroom and explain to Steve and Ive that you cannot put radio waves through metal," Apple engineer Phil Kearney said. "And it was not an easy explanation. Most of the designers are artists. The last science class they took was in eighth grade. But they have a lot of power at Apple. So they asked, 'Why can't we just make a little seam for the radio waves to escape through?' And you have to explain to them why you just can't."[30]

When it came to other hardware decisions, Jobs's exacting demands won out, often to the eventual benefit of the final product. The screen of the phone was originally supposed to be composed of the same plastic that iPod screens were made of. But after a day in Jobs's pocket, one prototype unit suffered from deep and permanent scratches thanks to his car keys. Jobs switched the screen from plastic to Gorilla Glass, even talking the glass maker Corning into converting an entire factory in Harrodsburg, Kentucky, just to produce the quantities Apple needed. This actually further complicated things for the hardware team, since the multitouch sensors now had to be embedded in glass, and glass was an entirely different proposition from embedding in plastic.

Other issues were solved by a clever combination of hardware and software. To make sure the screen turned off when a user pressed it to her face to answer a call, a proximity sensor was embedded. The problem of the phone accidentally turning on in a user's pocket was solved when a UI designer noticed the sliding lock and unlock mechanism on airplane bathroom doors. Thus, "slide to unlock" was born. Small but meaningful details were added as a result of the dogfooding feedback; details like a ringer switch to silence phone calls that came at inopportune times. The first person to actually receive a phone call on an iPhone was Andy Grignon. He was in a meeting and didn't recognize the caller's number, so he hit the ringer switch to ignore the call. "Instead of being this awesome Alexander Graham Bell moment," Grignon recalled that the first iPhone call was anticlimactic, "it was just like, 'Yeah, fuck it, go to voicemail.'"[31]

But the biggest headache, until late in the development period, remained the functionality of the software keyboard. The problem was finger size. If you tried to type, say, the letter "e," your finger might trigger a range of other letters instead. The solution, as ever, came from clever design. Apple engineers used artificial intelligence techniques to create an algorithm that would predict which letter a user might want to type next. For example, if someone types the letter "t," there is a very high probability that they will want to type "h" next. So, the letter "h" would, to the naked eye, look like it stayed the same size on the keyboard when, in fact, the "hit area" for the letter h would get bigger. After that, the "e" would likely be huge as a hit region. "The," after all, is a common word. This predictive typing algorithm saved the iPhone from repeating the failures of the Newton.

Even weeks and days before Apple was scheduled to announce the iPhone at the Macworld Conference in January 2007, the phone was still incredibly buggy. Demoing a half-baked product was not how Steve Jobs was used to doing things, but his hand was forced in this case. The fact that an Apple phone was coming was common knowledge. Reviewers, bloggers and reporters had whipped up an incredible frenzy of excitement over what they dubbed the "Jesus Phone." It *had* to debut.

Just after New Year's Day 2007, Apple took over the Moscone Center in San Francisco to host the iPhone launch event. A lone Apple employee was tasked with shepherding all twenty-four of the demo units in the trunk of his Acura, driving up from Apple headquarters in Cupertino, and delivering them to San Francisco. He was followed by a second car piloted by Apple security. The engineer wondered what would happen if he got into an accident and the demos were destroyed.

Jobs rehearsed his presentation for six solid days, but at the final hour, the team still couldn't get the phone to behave through an entire run-through. Sometimes it lost Internet connection. Sometimes the calls wouldn't go through. Sometimes the phone just shut down. In these moments, Jobs's notorious temper blazed to life. "It quickly got very uncomfortable," Andy Grignon said. "Very rarely did I see him become completely unglued. It happened. But mostly he just looked at you and very directly said in a very loud and stern voice, 'You are fucking up my company,' or, 'If we fail, it will be because of you.'"[32]

At the last minute, the engineers identified a "golden path," a specific set of demo actions that Jobs could perform in a specific order that afforded them the best chance of the phone making it through the presentation without a glitch. For example, Jobs could send an email and *then* surf the web, but if he reversed the order, the phone tended to crash. The engineers also masked the WiFi that Jobs would be using onstage so that audience members couldn't jump on the same network and possibly clog it up. AT&T brought in a portable cell tower to make sure Jobs would have a strong signal when he made his own first demo phone call. But, just to be on the safe side, the engineers hard-coded all the demo units to display five bars of cell strength, whether that happened to be true or not.

■

IT'S A SIGN OF THE technologically obsessed era we live in that Steve Jobs's Macworld keynote presentation on January 9, 2007, has gone down as a seminal moment in popular culture.

"This is a day I've been looking forward to for two and a half years," Jobs said somberly, walking across the width of the stage. "Every once in a while, a revolutionary product comes along that changes everything."

Apple executive Eddy Cue would say later: "It was the only event I took my wife and kids to because, as I told them, 'In your lifetime, this might be the biggest thing ever.' Because you could feel it. You just knew that this was huge."[33]

The words Jobs used to unveil the iPhone have become mythical:

> So . . . Three things: A widescreen iPod with touch controls. A revolutionary mobile phone. And a breakthrough Internet communications device. An iPod . . . a phone . . . and an Internet communicator . . . An iPod . . . a phone . . . are you getting it? These are not three separate devices. This is one device! And we are calling it iPhone.

Somehow, the demo went off without a hiccup. Watching the video now, as hundreds of millions have done on YouTube, Jobs is masterful, seemingly at the very height of his powers as a showman. You can feel him simultaneously stoking and feeding off the excitement emanating from the crowd. It is almost as if Jobs can't believe what he is demoing at the same time the audience can't believe what they're seeing.

The original iPhone that went on sale June 29, 2007, was based on the Purple 2 prototype, code named M68, with device number iPhone1,1. With more than a decade of perspective, perhaps the most remarkable thing about the first iPhone was that it was so completely, conceptually perfect, right out of the gate. Automobiles had to evolve for almost forty years until they settled into the standard configuration we are familiar with today. On their first attempt, the team at Apple managed to stumble upon the perfect form factor, the perfect incarnation of the modern smartphone. Smartphones had, of course, existed for several years previous to the iPhone, but the standard form of the smartphone as we know it today—no physical keyboard, a single slab of screen, a "black mirror" that is both a reflection of, and a conduit for all of our hopes and desires—they nailed it on the first try. And that's quite remarkable. There's a very good reason why, to this day, almost all smartphones essentially look like that first iPhone.

The iPhone, of course, solved the threat to Apple's iPod franchise by basically obsoleting the stand-alone MP3 player, just as it was designed to do. But what's often overlooked now is how important that "third" thing was that Jobs declared the iPhone to encompass at its core: an Internet communicator. Smartphones and PDAs had been gaining the ability to browse the web for years. But the iPhone had that comparatively enormous LCD screen that took up nearly the entire surface real estate of the device. And it had all of the multitouch advancements like pinch to zoom and double-tap to center on text. These were the things that made browsing the mobile web useful and enjoyable for the first time. Jobs himself would later say the miracle of mobile browsing was what truly made the first iPhone stand out. The iPhone delivered the "real" Internet like a "real" computer did. The iPhone finally made the mobile web a self-evident, useful feature. "It's the Internet in your pocket for the first time," Jobs said.[34]

The first iPhone, however, cannot actually lay claim to being the device that finally made the smartphone into the most successful computing device in history. Something people tend to forget about the first iPhone is how neutered it was. It was launched onto the nearly obsolete EDGE network. Cingular/AT&T was still in the process of building out its 3G network, so for that first-generation phone, users had to make do with snail-like data speeds. The first iPhone also lacked a GPS sensor, so even though you could use mobile maps in the first iPhone, the experience wasn't as seamless or accurate as it is today. The first iPhone couldn't shoot video, and didn't even have a front-facing camera, so the era of the "selfie" didn't come into being until the fourth generation of the iPhone, three years later.

■

THE BIGGEST REASON the first iPhone is not the iPhone of popular memory is that it didn't have the App Store. The first iPhone had the usual suite of PDA-like apps, a calendar, a notepad, a calculator, a clock, a stock ticker and a weather app, all designed by Apple. The only outside apps were the maps provided by Google and YouTube. There was no second screen to swipe to beyond the homescreen—because there were no other apps to put on the homescreen.

The original, App Store–less iPhone was very much Steve Jobs's platonic ideal of a closed and curated computing system, a perfect, hermetically sealed device. For several months after the iPhone's launch, Jobs was actually vocally opposed to the very idea of an app store, refusing to let outside developers infect his perfect creation. He told the *New York Times*: "You don't want your phone to be like a PC. The last thing you want is to have loaded three apps on your phone and then you go to make a call and it doesn't work anymore. These are more like iPods than they are like computers."[35]

But, in fact, Jobs was wrong about that. The iPhone very much *was* a computer. Back when the bake-off between the P1 and P2 models was happening, there was a simultaneous decision to be made in terms of what software would be used to run the device: a souped-up version of the iPod OS, or a scaled-down version of OS X, the OS that ran Apple's Mac computers. OS X came out the winner. Right out of the gate, the iPhone was, at least when it came to software architecture, a tiny but fully capable Mac. That meant that developers could write real, actual, full-blooded applications for the iPhone, if only Steve Jobs would allow them to do so.

In the end, the battle to do an app store was a replay of the argument over opening up iTunes to Windows users a few years earlier. Just as before, everyone inside Apple wanted to do it, and Jobs kept saying no. But in the end, just as with iTunes, the result was the same. Jobs finally caved, telling those who had been haranguing him, "Oh, hell, just go for it and leave me alone!"[36]

The iPhone App Store was launched in July 2008, alongside the second-generation iPhone 3G. As the former Apple employee Jean-Louis Gassée has said, "It was only then that the iPhone was truly finished, that it had all its basics, all its organs. It needed to grow, to muscle up, but it was complete as a child is complete."[37] In the first quarter the iPhone was on sale, Apple and AT&T sold about 1.5 million iPhones.[38] In the quarter after the App Store launched, Apple sold 6.89 million, exceeding 10 million total iPhone sales for the first time, and surpassing RIM's BlackBerry to become the best-selling smartphone in the United States.[39]

■

IT WAS THE APP STORE that inspired users to adopt smartphones and make them mainstream. Smartphone ownership in America went from 3% in 2007 when the iPhone was announced, to more than 80% a decade later.[40] At the time of this writing, the iPhone has sold over a billion units and Apple is the most valuable company in the world. Certainly, some of this stratospheric success was due to the hardware designs that Jony Ive came up with, which made each successive iPhone an object of lust and envy. We can also credit Steve Jobs's consummate showmanship for making the smartphone into the iconic device of the modern era. But more than anything else, we have to credit the App Store for turning the smartphone from a niche category that only appealed to early adopters and on-the-go professionals into a universal computer that appealed to everyone and their mother.

In a larger sense, the iPhone and the App Store were triumphs of software. "Software wrapped in a beautiful package," was how Steve Jobs liked to describe it.[41] Just as Bill Gates had intuited all the way back in the 1970s, software was the key differentiator. Software was what made mobile computers indispensable. "There's an app for that" was not just a clever marketing concept, it actually reflected how smartphones—via mobile apps—were able to subsume all of the lessons of the Internet Era. Getting the latest news, buying from Amazon or eBay, searching Google, looking up a fact on Wikipedia, listening to an unlimited selection of music (the promise of Napster), watching a YouTube video, streaming Netflix—every single miracle of the web revolution of the previous fifteen years found new life on the tiny computers in our pockets. Thanks to the triumph of software, the iPhone even allowed Apple to create a true platform, an ecosystem that the mobile computing world has to exist within. It was just what Marc Andreessen had dreamed of back at Netscape.

But if we're being entirely honest, there's one specific category of app that was crucial to the iPhone taking off when previous smartphones didn't.

One of the key launch apps on the first day the App Store went live? Facebook.

Social networks succeeded in making the Internet truly a personal experience. Smartphones, combined with social networks, took personal computing and made it almost intimate computing. Where would social media be without mobile computing, without smartphones: the perfect tools, always on hand to record and organize the ephemera of our daily lives? Would Facebook be at a billion users today if smartphones, in the example of iPhone, hadn't presented the perfect vehicle for social media consumption *and* production? And if not for the iPhone kick-starting the smartphone revolution, whither Snapchat? Or Twitter? Much less, Uber?

The argument could be made that social media finally broke through to the mainstream because smartphones went mainstream at the same time. And a complementary argument could be made in reverse: that the iPhone took off when other smartphones hadn't because it arrived on the scene just when Facebook was going parabolic.

Rather than too soon, the smartphone+social media represented a moment when two world-changing technologies arrived at just the right moment.

OUTRO

One of the true godfathers of the Internet was a man by the name of J. C. R. Licklider. In the 1950s, he worked at Bolt, Beranek & Newman, which would go on to build the computers that were connected to the first four nodes of the ARPANET. In the early 1960s, Licklider was the head of the Information Processing Techniques Office at ARPA, which would go on to fund the ARPANET. In 1963, he wrote the key internal paper that would plan for, and ultimately make the case for, the development of the ARPANET, the key precursor to today's Internet.

But like most computer scientists of his era, Licklider was also a theoretical visionary. In 1960, he wrote a paper called "Man-Computer Symbiosis," which is considered a fundamental text of modern computer science.

In 1960, as today, there were many who believed that true artificial intelligence was just around the corner. Licklider, however, put his money on cybernetics, the idea that man would meld with machine. In "Man-Computer Symbiosis," Licklider argued that thinking machines many orders of magnitude smarter than humans might arrive someday. They might even be inevitable. But in the meantime:

> There will nevertheless be a fairly long interim during which the main intellectual advances will be made by men and computers working together in intimate association.

> The hope is that, in not too many years, human brains and computing machines will be coupled together very tightly and that the resulting partnership will think as no human brain has ever thought and process data in a way not approached by the information-handling machines we know today.
>
> . . . Men will set the goals, formulate the hypotheses, determine the criteria, and perform the evaluations. Computing machines will do the routinizable work that must be done to prepare the way for insights and decisions in technical and scientific thinking. Preliminary analyses indicate that the symbiotic partnership will perform intellectual operations much more effectively than man alone can perform them.

At its core, the Internet Era represents that "fairly long interim" that Licklider envisioned, where humanity and computers came together in profound ways. First, we connected all the world's computers together. Then, we uploaded all of humanity's collected knowledge into the virtual space that networks created. Then, we made all of that knowledge searchable. We tied our commerce systems, our financial systems, even our media and information systems, to the network. We created a world where any good, any piece of media, any piece of art, any fact or thought, any idea or meme, is available, on call, for the instant gratification of any curiosity or desire. Over the course of a decade, we learned how to behave, and then to actually live with this new networked paradigm—to actually *exist in* this virtual environment. With social media, we connected ourselves together just as comprehensively as we had connected all the computers. And then, we started wearing actual supercomputers on our bodies, taking them with us at every waking moment of our days, to navigate, not only the intellectual, the social, but even the physical space of modern life. And we did all this unbidden, undirected, unplanned—almost as if we were following a biological impulse, guided by some unconscious evolutionary imperative.

When you see everyone around you hunched over the glowing

screens of their smartphones, you're seeing the fulfillment of the intimate association of man and machine that Licklider envisioned.

But are we better off? Are we truly thinking as no human brain has ever thought, just as Licklider supposed?

That's the open-ended question as the Internet Era continues.

ACKNOWLEDGMENTS

This book began as a podcast called the Internet History Podcast, so my first thanks go to the hundreds of people who have allowed me to interview them and share their stories of making the Internet Era happen. There are too many guests to thank in this limited space, but if you enjoyed this book, I highly recommend giving the podcast a try, to hear the stories that informed this project, and the details and anecdotes that didn't make it into this volume. My intention is that the podcast will continue for years to come, preserving oral histories of technology for posterity. Check it out at www.internethistorypodcast.com or on your podcast app of choice.

Thanks to three podcast guests in particular. Nancy Evans, thank you for kicking my butt and reminding me this project deserved to be a book. And thank you to Ben Slivka and Chris Fralic for seeing real value in the project at key moments. Thanks wholeheartedly to my agent, Kevin O'Connor, for believing this was a big, important book from day one, and also to my editor at Liveright, Katie Adams, for taking this project on and deftly ensuring that it wasn't a big, bloated book. Thanks to Fred Wiemer and Amy Medeiros for expert copy editing. Thanks to Phil Marino for being the first one to take a chance on this idea.

Thanks to Bill McManus, for getting my headspace to a place where this project was possible. Thanks to Angelita Sosa for taking care of my family and getting us to a place where this project was feasible. Thanks to Joel Lovell for key advice at the right time. Thanks to Chris Anderson and everyone at TED for making the

TED Residency happen, an amazing program that took this project to the next level. Special, loving thanks to all my fellow Residents, and especially Cyndi Stivers and Katrina Conanan, who run the Residency and are truly doing the Lord's work.

Thanks to the B. Altman Reference Desk at the Science, Industry and Business (SIBL) branch of the New York Public Library. People think everything is online and forever now, but don't believe it! Libraries are still vital. I would not have been able to source half of this book if not for the physical references preserved from the dot-com era and before. I also want to thank the Manistee, Michigan, public library for giving me refuge in which to write this book, for several summers in a row.

Finally, everyone thanks their spouse when they write a book, for enduring the distraction and absences (physical and mental) that writing a book requires. But truly, my wife, Lesa Rozmarek, deserves unending credit and gratitude for supporting a side project/hobby/distraction that has, unplanned and unexpectedly, become a second career. Lesa, your completely unshakable and bottomless belief in me, while unfathomable at times, is welcome and cherished. I love you.

NOTES

INTRO

1 Susannah Fox and Lee Rainie, "The Web at 25 in the U.S.: Part 1—How the Internet Has Woven Itself into American Life," last modified February 27, 2014, http://www.pewinternet.org/2014/02/27/part-1-how-the-internet-has-woven-itself-into-american-life/.
2 Google Answers, "Q: Personal Computer Penetration in US," posted July 28, 2004, http://answers.google.com/answers/threadview?id=380304.
3 Google Groups, posted August 6, 1991, https://groups.google.com/forum/#!msg/alt.hypertext/eCTkkOoWTAY/bJGhZyooXzkJ.

1. THE BIG BANG: THE MOSAIC WEB BROWSER AND NETSCAPE

1 Molly Baker, "Technology Investors Fall Head over Heels for Their New Love," *Wall Street Journal*, August 10, 1995.
2 William Stewart, "NSFNET—National Science Foundation Network," Living Internet, accessed August 18, 2016, http://www.livinginternet.com/i/ii_nsfnet.htm.
3 Internet History Podcast, *Episode 8: Aleks Totic, of Mosaic and Netscape*, March 16, 2014.
4 Internet History Podcast, *Episode 9: Jon Mittelhauser, Founding Engineer, Mosaic and Netscape*, March 27, 2014.
5 Patricia Sellers, "Don't Call Me SLACKER! Meet America's Top Talents Under 30. They Are Unorthodox, Rebellious, and a Challenge to Manage," *Fortune*, December 12, 1994.
6 Internet History Podcast, *Episode 10: Rob McCool, Founding Engineer, Mosaic and Netscape*, April 2, 2014.
7 Ibid.
8 John Naughton, *A Brief History of the Future: From Radio Days to Internet Years in a Lifetime* (Woodstock, NY: Overlook, 2000), 239.
9 "Html & Emacs," e-mail message, November 16, 1992, accessed August 18, 2016, from the Internet Archive Wayback Machine, http://web.archive.org/web/20021225141741/http://ksi.cpsc.ucalgary.ca/archives/WWW-TALK/www-talk-1992.messages/292.html.

10 George Gilder, "The Coming Software Shift," *Forbes ASAP*, August 28, 1995.
11 Internet History Podcast, *Episode 8: Aleks Totic, of Mosaic and Netscape.*
12 James Gillies and Robert Cailliau, *How the Web Was Born: The Story of the World Wide Web* (Oxford: Oxford University Press, 2000), 241.
13 Matthew Gray, "Web Growth Summary," Internet Statistics, Massachusetts Institute of Technology, accessed August 18, 2016, http://www.mit.edu/~mkgray/net/web-growth-summary.html.
14 Gillies and Cailliau, *How the Web Was Born*, 242.
15 Tom Steinert-Threlkeld, "Can You Work in Netscape Time?" *Fast Company*, October 31, 1995.
16 Robert Reid, *Architects of the Web: 1,000 Days That Built the Future of Business* (New York: John Wiley & Sons, 1997), 12.
17 Walter Isaacson, *The Innovators: How a Group of Hackers, Geniuses, and Geeks Created the Digital Revolution* (New York: Simon & Schuster, 2014), 418.
18 Gillies and Cailliau, *How the Web Was Born*, 242.
19 Gilder, "The Coming Software Shift."
20 Woods Wilton, "1994 Products of the Year," *Fortune*, December 12, 1994.
21 Reid, *Architects of the Web*, 17.
22 Internet History Podcast, *Episode 6: Mosaic and Internet Explorer Engineer, Chris Wilson*, March 10, 2014.
23 Gilder, "The Coming Software Shift."
24 John Markoff, "Business Technology; A Free and Simple Computer Link," *New York Times*, December 8, 1993, D5.
25 Internet History Podcast, *Episode 8: Aleks Totic, of Mosaic and Netscape.*
26 Jonathan Weber, "Computer Sales Suffered a Rare Drop Last Year," *Los Angeles Times*, January 21, 1992.
27 *Tim Ferriss Show*, "163: Marc Andreessen—Lessons, Predictions, and Recommendations from an Icon," https://tim.blog/2016/05/29/marc-andreessen/.
28 David A. Kaplan, *The Silicon Boys and Their Valley of Dreams* (New York: HarperCollins, 2000), 231.
29 Jim Clark and Owen Edwards, *Netscape Time: The Making of the Billion-Dollar Start-up That Took on Microsoft* (New York: St. Martin's, 1999), 32.
30 Adam Lashinsky, "Remembering Netscape: The Birth of the Web—July 25, 2005," *Fortune*, July 25, 2005, http://archive.fortune.com/magazines/fortune/fortune_archive/2005/07/25/8266639/index.htm.
31 Internet History Podcast, *Episode 9: Jon Mittelhauser, Founding Engineer, Mosaic and Netscape.*
32 Internet History Podcast, *Episode 8: Aleks Totic, of Mosaic and Netscape.*
33 Internet History Podcast, *Episode 10: Rob McCool, Founding Engineer, Mosaic and Netscape.*
34 Clark and Edwards, *Netscape Time*, 58.
35 Internet History Podcast, *Episode 8: Aleks Totic, of Mosaic and Netscape.*
36 Jon Mittelhauser, "[IAmA] Co-author of the First Widely Used Web Browser, an Early Owner/Evangelist/Investor for Tesla Motors, and the

Guy Who Ran the Launch of OnLive (a Reddit Trifecta?) AMAA," Reddit, posted July 8, 2011, accessed August 19, 2016, https://www.reddit.com/r/IAmA/comments/ik5mk/iama_coauthor_of_the_first_widely_used_web/.

37 Clark and Edwards, *Netscape Time*, 63.

38 Ibid.

39 Jamie Zawinski, "The Netscape Dorm," accessed August 19, 2016, from the Internet Archive Wayback Machine capture on February 8, 2010, https://web.archive.org/web/20100208023804/http://www.jwz.org/gruntle/nscpdorm.html.

40 Internet History Podcast, *Episode 8: Aleks Totic, of Mosaic and Netscape.*

41 Internet History Podcast, *Episode 5: Netscape and Mosaic Founding Engineer, Lou Montulli*, March 6, 2014.

42 Reid, *Architects of the Web*, 27.

43 Internet History Podcast, *Episode 10: Rob McCool, Founding Engineer, Mosaic and Netscape.*

44 Joshua Quittner and Michelle Slatalla, *Speeding the Net: The Inside Story of Netscape and How It Challenged Microsoft* (New York: Atlantic Monthly Press, 1998), 121.

45 Ibid.

46 "The 25 Most Intriguing People in '94," *People*, December 26, 1994–January 2, 1995.

47 Internet History Podcast, *Episode 10: Rob McCool, Founding Engineer, Mosaic and Netscape.*

48 Internet History Podcast, *Episode 5: Netscape and Mosaic Founding Engineer, Lou Montulli.*

49 Reid, *Architects of the Web*, 31.

50 Internet History Podcast, *Episode 9: Jon Mittelhauser, Founding Engineer, Mosaic and Netscape.*

51 Internet History Podcast, *Episode 8: Aleks Totic, of Mosaic and Netscape.*

52 Robert D. Hof, "From the Man Who Brought You Silicon Graphics . . ." *BusinessWeek*, October 24, 1994.

53 Quittner and Slatalla, *Speeding the Net*,174.

54 "Layout Engine Usage Share," Wikipedia, accessed August 19, 2016, https://en.wikipedia.org/wiki/File:Layout_engine_usage_share-2009-01-07.svg.

55 Steinert-Threlkeld, "Can You Work in Netscape Time?"

56 Lashinsky, "Remembering Netscape."

57 Naughton, *A Brief History of the Future: From Radio Days to Internet Years in a Lifetime*, 251.

58 Quittner and Slatalla, *Speeding the Net*, 203.

59 Internet History Podcast, *Episode 9: Jon Mittelhauser, Founding Engineer, Mosaic and Netscape.*

60 U.S. Department of Justice, *U.S. v. Microsoft: Proposed Findings of Facts*, https://www.justice.gov/sites/default/files/atr/legacy/2006/04/10/iii-b.pdf.

61 Lashinsky, "Remembering Netscape."

62 Ibid.

63 Internet History Podcast, *Episode 9: Jon Mittelhauser, Founding Engineer, Mosaic and Netscape.*

64 James Collins, "High Stakes Winners," *Time*, February 19, 1996.

65 Michael Lewis, *The New New Thing: A Silicon Valley Story* (New York: W. W. Norton, 2000), 74.
66 Michael A. Cusumano and David B. Yoffie, *Competing on Internet Time: Lessons from Netscape and Its Battle with Microsoft* (New York: Free Press, 1998), 10.
67 Jeff Pelline, "Netscape Playing Catch-up to Yahoo," CNET, March 30, 1998, accessed August 19, 2016, http://www.cnet.com/news/Netscape-playing-catch-up-to-yahoo/.
68 Cusumano and Yoffie, *Competing on Internet Time*, 31.
69 Bob Metcalfe, "Without Case of Vapors, Netscape's Tools Will Give Blackbird Reason to Squawk," *InfoWorld*, September 18, 1995.

2. BILL GATES "GETS" THE INTERNET: MICROSOFT AND INTERNET EXPLORER

1 Brent Schlender, "What Bill Gates Really Wants," *Fortune*, January 16, 1995.
2 Ibid.
3 Laurent Belsie and Scott Armstrong, "High Hopes and Hype Blaze Path for Information Superhighway," *Christian Science Monitor*, January 13, 1994.
4 Alan Stone, *How America Got On-line: Politics, Markets, and the Revolution in Telecommunications* (Armonk, NY: M. E. Sharpe, 1997), 196.
5 David Kline, "Align and Conquer," *Wired*, February 1, 1995.
6 Edmund L. Andrews, "Time Warner's 'Time Machine' for Future Video," *New York Times*, December 11, 1994.
7 L. J. Davis, *The Billionaire Shell Game: How Cable Baron John Malone and Assorted Corporate Titans Invented a Future Nobody Wanted* (New York: Doubleday, 1998), 221.
8 Ibid., 179.
9 Internet History Podcast, *Episode 88: How Microsoft Went Online, with Brad Silverberg*, November 2, 2015.
10 "Time 25," *Time*, June 17, 1996.
11 Paul Andrews, *How the Web Was Won: Microsoft from Windows to the Web; The Inside Story of How Bill Gates and His Band of Internet Idealists Transformed a Software Empire* (New York: Broadway, 1999), 63.
12 Ibid., 54.
13 J. Allard, "Windows: The Next Killer Application on the Internet," e-mail to Paul Maritz et al., January 25, 1994, di_killerapp_internetmemo.rtf.
14 Andrews, *How the Web Was Won*, 109.
15 James Wallace, *Overdrive: Bill Gates and the Race to Control Cyberspace* (New York: John Wiley, 1997), 183.
16 Andrews, *How the Web Was Won*, 116.
17 Joshua Quittner and Michelle Slatalla, *Speeding the Net: The Inside Story of Netscape and How It Challenged Microsoft* (New York: Atlantic Monthly Press, 1998), 192.
18 Kathy Rebello, "Inside Microsoft: The Untold Story of How the Internet Forced Bill Gates to Reverse Course," *BusinessWeek*, July 15, 1996.
19 Tom Steinert-Threlkeld, "Can You Work in Netscape Time?" *Fast Company*, October 31, 1995.

20 Internet History Podcast, *Episode 88, How Microsoft Went Online.*

21 Joshua Cooper Ramo, "Winner Take All: Microsoft v. Netscape," *Time*, September 16, 1996.

22 Internet History Podcast, *Episode 88, How Microsoft Went Online.*

23 Rebello, "Inside Microsoft."

24 Wallace, *Overdrive*, 9.

25 Charles Cooper, "Of Silicon Valley and Sominex," *PC Week*, June 5, 1996.

26 Rebello, "Inside Microsoft."

27 Gary Wolf, "Steve Jobs: The Next Insanely Great Thing," *Wired*, February 1, 1996.

3. AMERICA, ONLINE: AOL AND THE EARLY ONLINE SERVICES

1 Michael A. Banks, *On the Way to the Web: The Secret History of the Internet and Its Founders* (Berkeley, CA: Apress, 2008), 93.

2 Ibid., 144.

3 Robert D. Shapiro, "This Is Not Your Father's Prodigy," Wired.com, January 6, 1993.

4 Glenn Rifkin, "At Age 9, Prodigy On-Line Reboots," *New York Times*, November 7, 1993.

5 Paul M. Eng, "Prodigy Is in That Awkward Stage," *BusinessWeek*, February 13, 1995; available at https://www.bloomberg.com/news/articles/1995-02-12/prodigy-is-in-that-awkward-stage.

6 Mark Nollinger, "America, Online!" Wired.com, September 1, 1995.

7 Kara Swisher, *AOL.com: How Steve Case Beat Bill Gates, Nailed the Netheads, and Made Millions in the War for the Web* (New York: Times Business/Random House, 1998), 94.

8 Nollinger, "America, Online!"

9 Jeff Goodell, "The Fevered Rise of America Online," *Rolling Stone*, October 3, 1996.

10 America Online, Inc., "America Online, Inc. Passes 200,000 Household Mark," PR Newswire, October 27, 1992.

11 Harry McCracken, "A History of AOL, as Told in Its Own Old Press Releases," Technologizer, posted May 24, 2010.

12 Ibid.

13 Kara Swisher, *There Must Be a Pony in Here Somewhere: The AOL Time Warner Debacle and the Quest for a Digital Future* (New York: Crown, 2003), 39.

14 Swisher, *AOL.com*, 82.

15 Internet History Podcast, *Episode 27: She Gave the World a Billion AOL CDs—An Interview With Marketing Legend Jan Brandt*, August 11, 2014.

16 "What Was the Conversion Rate of AOL CDs in the 1990s?" Quora, answered December 28, 2010, https://www.quora.com/What-was-the-conversion-rate-of-AOL-CDs-in-the-1990s/answer/Jan-Brandt.

17 Internet History Podcast, *Episode 27: She Gave the World a Billion AOL CDs.*

18 "How Much Did It Cost AOL to Distribute All Those CDs Back in the 1990s?" Quora, answered December 27, 2010, https://www.quora.com

/How-much-did-it-cost-AOL-to-distribute-all-those-CDs-back-in-the-1990s/answer/Jan-Brandt.

19 Internet History Podcast, *Episode 27: She Gave the World a Billion AOL CDs.*

20 Swisher, *AOL.com*, 102.

21 Ibid., 103.

22 Nollinger, "America, Online!"

23 Swisher, *AOL.com*, 124.

24 Nollinger, "America, Online!"

25 Swisher, *AOL.com*, 128.

26 Gene Koprowski, "AOL CEO Steve Case," *Forbes ASAP*, October 7, 1996.

27 Ibid.

28 Nollinger, "America, Online!"

29 David Carlson, "The Online Timeline, 1990–94," accessed January 31, 2018, http://iml.jou.ufl.edu/carlson/1990s.shtml.

30 Nollinger, "America, Online!"

31 Ibid.

32 Frank Rose, "Keyword: Context," *Wired*, December 1, 1996.

33 Amy Cortese and Amy Barrett, "The Online World of Steve Case," *BusinessWeek*, April 15, 1996; available online at https://www.bloomberg.com/news/articles/1996-04-14/the-online-world-of-steve-case.

34 Peter Coy, "Has the Net Finally Reached the Wall? America Online's Crash May Portend Constant Crises Unless the Internet Is Revamped," *BusinessWeek*, August 26, 1996.

35 Rose, "Keyword: Context."

36 Internet History Podcast, *Episode 27: She Gave the World a Billion AOL CDs.*

37 Swisher, *AOL.com*, 206.

38 Rose, "Keyword: Context."

39 Swisher, *AOL.com*, 206.

40 Ibid., 208.

41 Internet History Podcast, *Episode 27: She Gave the World a Billion AOL CDs.*

42 Swisher, *AOL.com*, 275.

4. BIG MEDIA'S BIG WEB ADVENTURE: PATHFINDER, *HOTWIRED* AND ADS

1 Chris Dixon, "The Next Big Thing Will Start Out Looking Like a Toy," *cdixon blog*, January 3, 2010, accessed February 9, 2018, http://cdixon.org/2010/01/03/the-next-big-thing-will-start-out-looking-like-a-toy/.

2 "The Web Back in 1996–1997," posted September 16, 2008, http://royal.pingdom.com/2008/09/16/the-web-in-1996-1997/.

3 Ibid.

4 Tim O'Reilly, "SLAC Symposium on the Early Web," posted November 26, 2001, https://web.archive.org/web/20141216174754/http://archive.oreilly.com/lpt/wlg/907.

5 William Glaberson, "In San Jose, Knight-Ridder Tests a Newspaper Frontier," *New York Times*, February 7, 1994.

6 Michael Shapiro, "The Newspaper That Almost Seized the Future," *Columbia Journalism Review*, November 2011; available online at https://archives.cjr.org/feature/the_newspaper_that_almost_seized_the_future.php.
7 Ken Auletta, *The Highwaymen: Warriors of the Information Superhighway* (New York: Random House, 1997), 315.
8 Jane Hodges, "Pathfinder Readies for Year Two," *Ad Age*, October 23, 1995.
9 Ibid.
10 Alec Klein, *Stealing Time: Steve Case, Jerry Levin, and the Collapse of AOL Time Warner* (New York: Simon & Schuster, 2003), loc. 1262, Kindle.
11 James Ledbetter, "The End of the Path?" *Industry Standard*, October 26, 1998.
12 Kara Swisher, *There Must Be a Pony in Here Somewhere: The AOL Time Warner Debacle and the Quest for a Digital Future* (New York: Crown, 2003), 91.
13 Ledbetter, "The End of the Path?"
14 Internet History Podcast, *Episode 33: HotWired CEO Andrew Anker*, September 22, 2014.
15 Ibid.
16 Rick Tetzeli, "The Internet and Your Business," *Fortune*, March 7, 1994.
17 Internet History Podcast, *Episode 35: Joe McCambley Discusses Advertising and the First Banner Ads*, October 6, 2014.
18 Internet History Podcast, *Episode 38: An Oral History of the Web's First Banner Ads*, October 27, 2014.
19 Internet History Podcast, *Episode 13: Co-Designer of the First Banner Ad, Co-Founder of Razorfish, Craig Kanarick*, April 17, 2014.
20 Ibid.
21 Ibid.
22 Internet History Podcast, *Episode 35, Joe McCambley Discusses Advertising and the First Banner Ads.*
23 Internet History Podcast, *Episode 33: HotWired CEO Andrew Anker.*
24 Lou Montulli, "The Reasoning behind Web Cookies," *The Irregular Musings of Lou Montulli*, May 14, 2013, from the Internet Archive Wayback Machine capture on December 27, 2013, https://web.archive.org/web/20131227064455/http://www.montulli-blog.com/2013/05/the-reasoning-behind-web-cookies.html.
25 "Interactive Ad Firms Grow on the Web," *Upside*, September 1996, 50.
26 Constance Loizos, "Feeling the Burn," *Red Herring*, April 1998.
27 George Slefo, "Digital Ad Spending Surges to Record High as Mobile and Social Grow More Than 50%," *Ad Age*, April 21, 2016.

5. HELLO, WORLD: THE EARLY SEARCH ENGINES AND YAHOO

1 Matthew Gray, "Web Growth Summary," 1996, accessed January 31, 2018, http://stuff.mit.edu/people/mkgray/net/web-growth-summary.html.
2 Cybertelecom, "History of DNS," accessed January 31, 2018, http://www.cybertelecom.org/dns/history.htm.
3 Jerry Yang and David Filo, *Yahoo! Unplugged: Your Discovery Guide to the Web* (New York: John Wiley, 1995), 198 and 240.

4 Randall E. Stross, "How Yahoo! Won the Search Wars," *Fortune*, March 2, 1998.
5 Ibid.
6 Ibid.
7 Ibid.
8 Brent Schlender, "How a Virtuoso Plays the Web," *Fortune*, March 6, 2000.
9 Internet History Podcast, *Episode 21: Yahoo Employee #3, Tim Brady*, June 16, 2014.
10 Robert Reid, *Architects of the Web: 1,000 Days That Built the Future of Business* (New York: John Wiley, 1997), 254.
11 Karen Angel, *Inside Yahoo!: Reinvention and the Road Ahead* (New York: John Wiley, 2002), 18.
12 Ibid.
13 David A. Kaplan, *The Silicon Boys and Their Valley of Dreams* (New York: HarperCollins, 2000), 310.
14 Ibid., 312.
15 Angel, *Inside Yahoo!*, 25.
16 Reid, *Architects of the Web*, 267.
17 Kaplan, *The Silicon Boys and Their Valley of Dreams*, 305.
18 Internet History Podcast, *Episode 78: Yahoo's Master Brand Builder, Karen Edwards*, August 24, 2015.
19 Linda Himelstein, Heather Green, Richard Siklos, and Catherine Yang, "Yahoo! The Company, the Strategy, the Stock," *BusinessWeek*, August 27, 1998.
20 Bernhard Warner, "Your Ad Here," *Industry Standard*, September 27, 1999.
21 Janice Maloney, "Yahoo: Still Searching for Profits on the Internet," *Fortune*, May 1996.
22 Reid, *Architects of the Web*, 265.
23 Ibid., 262.
24 Angel, *Inside Yahoo!*, 57.
25 Reid, *Architects of the Web*, 264.
26 Stross, "How Yahoo! Won the Search Wars."
27 Angel, *Inside Yahoo!*, 57.
28 Stross, "How Yahoo! Won the Search Wars."
29 Jeff Pelline, "Netscape Revenue Up 114%," CNET, April 23, 1997.
30 Securities and Exchange Commission, Yahoo! Inc. Form 10-Q, accessed January 31, 2018, http://www.sec.gov/Archives/edgar/data/1011006/00009 12057-96-017646.txt.
31 Angel, *Inside Yahoo!*, 45.
32 Kim Cleland, "A Gaggle of Web Guides Vies for Ads; Yahoo Directory Opens to Sponsorship Deals as Competition Grows," *Ad Age*, April 17, 1995.
33 Internet History Podcast, *Episode 21: Yahoo Employee #3, Tim Brady*.
34 Reid, *Architects of the Web*, 262.
35 Angel, *Inside Yahoo!*, 73.
36 Ibid., 81.
37 Ibid., 87.
38 Ibid., 98.

6. GET BIG FAST: AMAZON.COM AND THE BIRTH OF ECOMMERCE

1 Brad Stone, *The Everything Store: Jeff Bezos and the Age of Amazon* (New York: Little, Brown, 2014), 25–26.

2 Robert Spector, *Amazon.com: Get Big Fast* (New York: HarperBusiness, 2000), 25.

3 Stone, *The Everything Store*, 26.

4 Ibid., 25.

5 Jeff Bezos, interview by Academy of Achievement, May 4, 2001, http://www.achievement.org/achiever/jeffrey-p-bezos/#interview.

6 Jeff Bezos, "A Bookstore by Any Other Name" (lecture, Commonwealth Club of California, July 27, 1998).

7 Michael Dunlop, "10 World Famous Companies That Started in Garages," Retire@21, accessed January 31, 2018, http://www.retireat21.com/blog/10-companies-started-garages.

8 Internet History Podcast, *Episode 50: Amazon's Technical Co-Founder and Employee #1, Shel Kaphan*, February 1, 2015.

9 Stone, *The Everything Store*, 35.

10 Spector, *Amazon.com: Get Big Fast*, 52.

11 Stone, *The Everything Store*, 38.

12 David Sheff, "The Playboy Interview: Jeff Bezos," *Playboy*, February 1, 2000.

13 Internet History Podcast, *Episode 50: Amazon's Technical Co-Founder and Employee #1, Shel Kaphan.*

14 Ibid.

15 Spector, *Amazon.com: Get Big Fast*, 73.

16 Ibid., 85; Sheff, "The Playboy Interview: Jeff Bezos."

17 Stone, *The Everything Store*, 41.

18 Spector, *Amazon.com: Get Big Fast*, 93.

19 G. Bruce Knecht, "Wall Street Whiz Finds Niche Selling Books on the Internet," *Wall Street Journal*, May 16, 1996.

20 Stone, *The Everything Store*, 48.

21 Ibid.

22 James Romenesko, "The Height of Online Success: Tiny Amazon.com Squares Off Against Industry Giant Barnes & Noble," *St. Paul Pioneer Press*, July 21, 1997, 6E.

23 Spector, *Amazon.com: Get Big Fast*, 114.

24 Ibid., 124.

25 Robert Spector, "Yesterday's Goliath, Today's David," *Wall Street Journal*, June 25, 2011.

26 Randall E. Stross, "Why Barnes & Noble May Crush Amazon," *Fortune*, September 29, 1997.

27 Spector, *Amazon.com: Get Big Fast*, 168.

28 Romenesko, "The Height of Online Success."

29 Sheff, "The Playboy Interview: Jeff Bezos."

30 Stone, *The Everything Store*, 59.

31 William C. Taylor, "Who's Writing the Book on Web Business?" *Fast Company*, October–November 1996.
32 Stone, *The Everything Store*, 54.
33 Pankaj Ghemawat, *Leadership Online (B): Barnes & Noble vs. Amazon.com in 2005*, Harvard Business School Case Study 9-705-492 (Boston: Harvard Business School, 2006).
34 Morris Rosenthal, "Book Sales Statistics," Foner Books, accessed January 31, 2018, http://www.fonerbooks.com/booksale.htm.
35 "Amazon History and Timeline," accessed January 31, 2018, http://phx.corporate-ir.net/phoenix.zhtml?p=irol-corporateTimeline_pf&c=176060.
36 Stone, *The Everything Store*, 67.
37 Spector, *Amazon.com: Get Big Fast*, 161.

7. TRUSTING STRANGERS: EBAY, COMMUNITY SITES AND PORTALS

1 Adam Cohen, *The Perfect Store: Inside eBay* (New York: Little, Brown, 2002), 20.
2 Ibid., 22.
3 Google Groups, posted September 12, 1995, https://groups.google.com/forum/#!msg/misc.forsale.non-computer/DxxiU7FQp8Q/8ncYwB2DDEAJ.
4 Cohen, *The Perfect Store*, 25.
5 Ibid., 29.
6 Ibid., 44.
7 Ibid., 55.
8 Ibid., 59.
9 Ibid., 48.
10 Ibid., 46.
11 Joshua Cooper Ramo, "The Fast-Moving Internet Economy Has a Jungle of Competitors . . . and Here's the King," *Time*, December 27, 1999.
12 Cohen, *The Perfect Store*, 83.
13 Ibid., 57.
14 Ibid., 64.
15 Ibid., 79.
16 Ibid., 110.
17 Casey Hait and Stephen Weiss, *Digital Hustlers: Living Large and Falling Hard in Silicon Alley* (New York: HarperCollins, 2001), 47.
18 Ibid., 46.
19 *Silicon Alley Reporter*, no. 16 (Summer 1998), 38.
20 Hait and Weiss, *Digital Hustlers*, 115.
21 Internet History Podcast, *Episode 91: Co-Founder of Feed Magazine, Stefanie Syman*, December 7, 2015.
22 Internet History Podcast, *Episode 107: Founder of Marketwatch, Larry Kramer @lkramer*, May 22, 2016.
23 David Plotz, "A Slate Timeline," *Slate*, June 19, 2006; accessed January 31, 2018, http://www.slate.com/articles/news_and_politics/slates_10th_anniversary/2006/06/a_slate_timeline.html.
24 Stephen P. Bradley and Erin E. Sullivan, *AOL Time Warner, Inc.*, Harvard Business School Case Study 9-702-421, June 23, 2005.

25 Internet History Podcast, *Episode 62: iVillage Co-Founder Nancy Evans*, April 27, 2015.
26 Ibid.
27 Internet History Podcast, *Episode 64: GeoCities Founder David Bohnett*, May 11, 2015.
28 Eric Ransdell, "Broadcast.com Boosts Its Signal," *Fast Company*, August 1998.
29 Mike Sager, "The Billionaire," *Esquire*, April 1, 2000.
30 Ransdell, "Broadcast.com Boosts Its Signal."
31 Po Bronson, *The Nudist on the Late Shift: And Other True Tales of Silicon Valley* (New York: Broadway, 2000), 78.
32 Po Bronson, "Hotmale," *Wired*, December 1, 1998.
33 Karen Angel, *Inside Yahoo!: Reinvention and the Road Ahead* (New York: John Wiley, 2002), 86.
34 Ibid., 89.
35 PR Newswire, "April Internet Ratings from Nielsen//NetRatings," May 11, 1999, https://www.thefreelibrary.com/April+Internet+Ratings+From+Nielsen%2F%2FNetRatings.-a054609261.
36 Gordon Gould, "Search and Destroy," *Silicon Alley Reporter*, no. 16 (Summer 1998).
37 Jim Evans, "Portals in a Storm," *Industry Standard*, December 28, 1998–January 4, 1999.
38 Angel, *Inside Yahoo!*, 93.
39 Ibid., 131.
40 Ibid., 80.
41 Michael Krantz, "Start Your Engines: Excite and Yahoo, the Two Leading Web-Search Sites, Race to Remake Themselves into Portals," *Time*, April 20, 1998.
42 Linda Himelstein, Heather Green, Richard Siklos, and Catherine Yang, "Yahoo! The Company, the Strategy, the Stock," *BusinessWeek*, August 27, 1998.

8. BLOWING BUBBLES: THE DOT-COM ERA

1 John Cassidy, *Dot.Con: How America Lost Its Mind and Money in the Internet Era* (New York: HarperCollins, 2002), 28.
2 John Kenneth Galbraith, *A Short History of Financial Euphoria* (New York: Whittle Books in association with Viking, 1993), 87.
3 Internet History Podcast, *Episode 67: Journalist Maggie Mahar Discusses the Dot-com Bubble*, June 1, 2015.
4 Roger Lowenstein, *Origins of the Crash: The Great Bubble and Its Undoing* (New York: Penguin Press, 2004), 103.
5 Cassidy, *Dot.Con*, 107.
6 Charles Fishman, "The Revolution Will Be Televised (on CNBC)," *Fast Company*, June 2000.
7 Internet History Podcast, *Episode 67: Journalist Maggie Mahar Discusses the Dot-com Bubble.*
8 Andy Serwar, "A Nation of Traders," *Fortune*, October 11, 1999.
9 Maggie Mahar, *Bull! A History of the Boom and Bust, 1982–2004* (New York: HarperBusiness, 2003), 257.

10 Cassidy, *Dot.Con*, 200.
11 Ibid., 200.
12 Ibid., 201.
13 Mahar, *Bull!*, 292.
14 Ibid., xviii.
15 Joseph Nocera and Tyler Maroney, "Do You Believe? How Yahoo! Became a Blue Chip," *Fortune*, June 7, 1999.
16 Cassidy, *Dot.Con*, 162.
17 Jim Rohwer, "The Numbers Game," *Fortune*, November 22, 1999.
18 Mahar, *Bull!*, 263.
19 Federal Reserve Board, "Remarks by Chairman Alan Greenspan," December 5, 1996, https://www.federalreserve.gov/boarddocs/speeches/1996/19961205.htm.
20 Sebastian Mallaby, *The Man Who Knew: The Life and Times of Alan Greenspan* (New York: Penguin, 2016), 741.
21 Justin Martin, *Greenspan: The Man Behind the Money* (Cambridge, MA: Perseus, 2000), 226.
22 Mahar, *Bull!*, 6.
23 Ibid., 170.
24 Amy Kover, "Dot-com Time Bomb on Madison Avenue," *Fortune*, December 6, 1999.
25 Bethany McLean, "More Than Just Dot-coms," *Fortune*, December 6, 1999; Anthony B. Perkins and Michael C. Perkins, *The Internet Bubble: Inside the Overvalued World of High-Tech Stocks—And What You Need to Know to Avoid the Coming Shakeout* (New York: HarperBusiness, 1999), 6.
26 Brent Goldfarb, Michael Pfarrer, and David Kirsch, "Searching for Ghosts: Business Survival, Unmeasured Entrepreneurial Activity and Private Equity Investment in the Dot-com Era" (working paper RHS-06-027, Social Science Research Network, Rochester, SSRN-id929845, 2005, accessed March 26, 2017; downloadable at http:// papers.ssrn.com/ abstract = 825687).
27 John Cassidy, "Striking It Rich: The Rise and Fall of Popular Capitalism," *New Yorker*, January 14, 2002.
28 Mark Gimein, "Around the Globe, Net Stock Mania," *Industry Standard*, December 28, 1998–January 4, 1999.

9. IRRATIONAL EXUBERANCE: THE DOT-COM BUBBLE

1 Peter Elkind, "The Hype Is Big, Really Big, at Priceline," *Fortune*, September 6, 1999.
2 David Noonan, "Price Is Right," *Industry Standard*, December 28, 1998–January 4, 1999.
3 Ibid.
4 Ibid.
5 Elkind, "The Hype Is Big, Really Big, at Priceline."
6 Ibid.
7 Todd Woody, "Idea Man," *Industry Standard*, June 28, 1999.
8 Randall E. Stross, *eBoys: The First Inside Account of Venture Capitalists at Work* (New York: Ballantine, 2001), 120; Woody, "Idea Man."

9 Woody, "Idea Man."
10 Elkind, "The Hype Is Big, Really Big, at Priceline."
11 Ibid.
12 Dyan Machan, "An Edison for a New Age," *Forbes*, May 17, 1999.
13 Theta Pavis, "Toys 'R' Online," *Digital Coast Reporter*, no. 3 (October 1998).
14 Miguel Helft, "Uncle of the Board," *Industry Standard*, December 27, 1999–January 3, 2000.
15 Jacob Ward, "EToys 'R' Us?" *Industry Standard*, May 31–June 7, 1999.
16 Omar Merlo, *Pets.com Inc.: Rise and Decline of a Pet Supply Retailer*, Harvard Business Review Case Study 909A21, September 15, 2009.
17 Tim Clark, "Amazon Invests in Online Pet Store," *CNET News*, March 29, 1999, http://news.cnet.com/Amazon-invests-in-online-pet-store/2100-1017_3-223621.html.
18 Philip J. Kaplan, *F'd Companies: Spectacular Dot-com Flameouts* (New York: Simon & Schuster, 2002), 16.
19 Ibid., 21.
20 Securities and Exchange Commission, Webvan Group, Inc., Form 424B1, Prospectus filed with SEC, accessed February 9, 2018, https://www.sec.gov/Archives/edgar/data/1092657/0000891618990049140/000891618-99-004914.txt; Stross, *eBoys*, 36; Randall Stross, "Only a Bold Gamble Can Save Webvan Now," *Wall Street Journal*, February 2, 2001; and Scott Simon, "Profile: Online Shopping with Webvan," *Weekend Edition* (National Public Radio), October 9, 1999.
21 Linda Himelstein, "Louis H. Borders," *BusinessWeek*, September 27, 1999; and Himelstein, "Can You Sell Groceries Like Books?" *BusinessWeek*, August 26, 1999.
22 Andrew McAfee and Mona Ashiya, *Webvan*, Harvard Business Review Case Study 9-602-037, February 14, 2002.
23 Ibid.
24 Himelstein, "Can You Sell Groceries Like Books?"
25 Carolyn Said, "Online Beats In Line / Buying Groceries on the Web Takes the Hassle out of Shopping," *San Francisco Chronicle*, July 22, 1999.
26 Stross, "Only a Bold Gamble Can Save Webvan Now."
27 Rusty Weston, "Return of the Milkman," *Upside*, April 1, 2000.
28 McAfee and Ashiya, *Webvan*.
29 Securities and Exchange Commission, Webvan Group, Inc., Form 10-Q, accessed February 9, 2018, https://www.sec.gov/Archives/edgar/data/1092657/000089161800002826/0000891618-00-002826.txt; and Kara Swisher, "Webvan Needs Fresh Ideas to Help Bring Home Bacon," *Wall Street Journal*, October 2, 2000.
30 Saul Hansell, "Some Hard Lessons for Online Grocer," *New York Times*, February 19, 2001.
31 Arlene Weintraub and Robert D. Hof, "For Online Pet Stores, It's Dog-Eat-Dog," *BusinessWeek*, March 6, 2000.
32 Laurie Freeman, "Pets.com Socks It to Competitors," *Ad Age*, November 29, 1999, via Factiva, accessed November 17, 2008.
33 "Death of a Spokespup," *Adweek*, New England edition, December 2000.

34 Weintraub and Hof, "For Online Pet Stores, It's Dog-Eat-Dog."
35 Mike Tarsala, "Pets.com Killed by Sock Puppet," MarketWatch.com, November 8, 2000, https://www.marketwatch.com/story/sock-puppet-kills-petscom.
36 Freeman, "Pets.com Socks It to Competitors"; Tarsala, "Pets.com Killed by Sock Puppet."
37 Thomas Eisenmann, "Petstore.com," HBS No. 801-044 (Boston: Harvard Business School Publishing, 2000), p. 9.
38 Stross, *eBoys*, 116
39 Julia Flynn, "Gap Exists Between Entrepreneurship in Europe, North America, Study Shows," *Wall Street Journal*, July 2, 1999.
40 Stross, *eBoys*, 63.
41 Anthony B. Perkins and Michael C. Perkins, *The Internet Bubble: Inside the Overvalued World of High-Tech Stocks—And What You Need to Know to Avoid the Coming Shakeout* (New York: HarperBusiness, 1999), 38.
42 John Cassidy, *Dot.Con: How America Lost Its Mind and Money in the Internet Era* (New York: HarperCollins, 2002), 237.
43 Stross, *eBoys*, xvii.
44 Roger Lowenstein, *Origins of the Crash: The Great Bubble and Its Undoing* (New York: Penguin Press, 2004), 101, 112.
45 "Financial Spotlight: Net IPOs Lose Their Luster," *Industry Standard*, June 28, 1999.
46 Lowenstein, *Origins of the Crash*, 125.
47 "Yahoo! Buys GeoCities," CNNMoney, January 28, 1999. http://money.cnn.com/1999/01/28/technology/yahoo_a/.
48 Stephan Paternot, *A Very Public Offering: A Rebel's Story of Business Excess, Success, and Reckoning* (New York: J Wiley, 2001), 172.
49 Perkins and Perkins, *The Internet Bubble* (New York: HarperBusiness, 1999), 21.
50 "Excite@Home Buys Online Greeting Card Site for $780 Million," CNET, January 2, 2002, https://www.cnet.com/news/excitehome-buys-online-greeting-card-site-for-780-million/.
51 "Excite@Home to Acquire Bluemountain," *New York Times*, October 26, 1999, http://www.nytimes.com/1999/10/26/business/excite-home-to-acquire-bluemountain.html.
52 "Will K-Tel's Stock Fizzle?" CNET, June 10, 1998, https://www.cnet.com/news/will-k-tels-stock-fizzle/.
53 Ernst Malmsten, *Boo Hoo: A Dot.com Story from Concept to Catastrophe* (London: Random House Business Books, 2001), 111.
54 Kaplan, *F'd Companies*, 24.
55 Ibid., 34.
56 Ibid., 38.
57 "Pixelon.com Announces iBash '99," InterActive Agency, Inc., October 27, 1999, http://www.alanwallace.com/iagency/public_relations/archives/1999/pixelon.10.27.99.html.
58 David Kirkpatrick, "Suddenly Pseudo," *New York*, December 20, 1999, accessed February 1, 2018, at http://nymag.com/nymetro/news/media/internet/1703/.

59 Casey Hait and Stephen Weiss, *Digital Hustlers: Living Large and Falling Hard in Silicon Alley* (New York: HarperCollins, 2001), 267.
60 *Wired* Staff, "Steaming Video," *Wired*, November 1, 2000.
61 Hait and Weiss, *Digital Hustlers*, 240.
62 Corrie Driebusch, "Drkoop.com Epitomized Hype of Tech Boom and Bust," *Wall Street Journal*, April 26, 2015.
63 James Ledbetter, "The Final Frontier for Lou Dobbs?" *Industry Standard*, June 21, 1999.
64 Kirin Kalia, "A Giant Leap for Web-Kind," *Silicon Alley Reporter*, no. 28 (1999).
65 Gail Shister, "Sam Donaldson Enjoys Internet Interview Show," *Chicago Tribune*, January 1, 2000.
66 Joshua Cooper Ramo, "The Fast-Moving Internet Economy Has a Jungle of Competitors . . . and Here's the King," *Time*, December 27, 1999.
67 James Kelly, "That Man in the Cardboard Box," *Time*, December 27, 1999.
68 Joshua Quittner, "An Eye on the Future," *Time*, December 27, 1999.
69 David Kirkpatrick, "Is Net Investing a Sucker's Game?" *Fortune*, October 11, 1999.
70 Justin Fox, "Net Stock Rules: Masters of a Parallel Universe," *Fortune*, June 7, 1999.
71 Jacqueline Doherty, "Amazon.bomb," *Barron's*, May 31, 1999.
72 Brad Stone, *The Everything Store: Jeff Bezos and the Age of Amazon* (New York: Little, Brown, 2014), 100.
73 Ibid., 101.

10. POP!: NETSCAPE VS. MICROSOFT, AOL + TIME WARNER AND THE NUCLEAR WINTER

1 David B. Yoffie and Mary Kwak, "The Browser Wars, 1994–1998," Harvard Business School Case 798-094 (June 1998), 9.
2 Kara Swisher, "After a Life at Warp Speed, Netscape Quickly Logs Off," *Wall Street Journal*, November 25, 1998.
3 David Yoffie and Michael A. Cusumano, *Competing on Internet Time: Lessons from Netscape and Its Battle with Microsoft* (New York: Free Press, 1998), 9.
4 Ibid., 9.
5 Ibid., 33.
6 Ibid., 38.
7 Yoffie and Kwak, "The Browser Wars," 9; Eric Nee, "Up for Grabs?" *Fortune*, February 23, 1998.
8 Yoffie and Kwak, "The Browser Wars," 14.
9 Kenneth S. Corts and Deborah Freier, "A Brief History of the Browser Wars," Harvard Business School Case 9-703-517 (2003), 6.
10 "Netscape Breaks Free," *Economist*, March 28, 1998.
11 Nee, "Up for Grabs?"
12 John Heilemann, *Pride Before the Fall: The Trials of Bill Gates and the End of the Microsoft Era* (New York: HarperCollins, 2001), 199.
13 Ken Auletta, *World War 3.0: Microsoft and Its Enemies* (New York: Random House, 2001), 362.

14 Charles Arthur, *Digital Wars: Apple, Google, Microsoft and the Battle for the Internet* (Philadelphia: Kogan, 2012), 22.
15 "List of Public Corporations by Market Capitalization," Wikipedia, last modified January 23, 2018, https://en.wikipedia.org/wiki/List_of_public_corporations_by_market_capitalization.
16 Joe Steinbring, "How Many Personal Computers Are Sold per Year?" accessed February 1, 2018, https://steinbring.net/2011/how-many-personal-computers-are-sold-per-year/.
17 U.S. Census Bureau, "Computer and Internet Use in the United States: Population Characteristics," issued May 2013, http://www.census.gov/prod/2013pubs/p20-569.pdf.
18 Internet History Podcast, *Episode 8: Aleks Totic, of Mosaic and Netscape.*
19 Auletta, *World War 3.0*, 197.
20 Jared Sandberg, "WorldCom Agrees to Acquire CompuServe for $1.2 Billion," *Wall Street Journal*, September 8, 1997.
21 Nina Munk, *Fools Rush In: Steve Case, Jerry Levin, and the Unmaking of AOL Time Warner* (New York: HarperCollins, 2004), 118.
22 Marc Gunther, "AOL: The Future King of Advertising?" *Fortune*, October 11, 1999.
23 Munk, *Fools Rush In*, 118.
24 Marc Gunther, Liz Smith, and Wilton Woods, "The Internet Is Mr. Case's Neighborhood," *Fortune*, March 30, 1998, accessed February 1, 2018, http://archive.fortune.com/magazines/fortune/fortune_archive/1998/03/30/240097/index.htm.
25 Gunther, "AOL: The Future King of Advertising?"
26 "AOL, Drkoop.com Partner," CNNMoney, July 6, 1999, accessed February 1, 2018, http://money.cnn.com/1999/07/06/technology/aol/.
27 Kara Swisher, *There Must Be a Pony in Here Somewhere: The AOL Time Warner Debacle and the Quest for a Digital Future* (New York: Crown, 2003), 62.
28 Ibid., 109, 117.
29 Munk, *Fools Rush In*, 106.
30 Gary Rivlin, "AOL's Rough Riders," *Industry Standard*, October 30, 2000.
31 Munk, *Fools Rush In*, 153.
32 Justin Fox, "Net Stock Rules: Masters of a Parallel Universe," *Fortune*, June 7, 1999; Swisher, *There Must Be a Pony in Here Somewhere*, 119.
33 Munk, *Fools Rush In*, 118, 123.
34 Swisher, *There Must Be a Pony in Here Somewhere*, 128.
35 Munk, *Fools Rush In*, 125.
36 Swisher, *There Must Be a Pony in Here Somewhere*, 141.
37 Ibid., 154.
38 Ibid., 141.
39 Munk, *Fools Rush In*, 118.
40 Daniel Okrent, Maryanne Murray Buechner, Adam Cohen, Emily Mitchell, Michael Krantz, and Chris Taylor, "Happily Ever After?" *Time*, January 24, 2000.
41 Ibid.

42 Swisher, *There Must Be a Pony in Here Somewhere*, 155.
43 John Cassidy, *Dot.Con: How America Lost Its Mind and Money in the Internet Era* (New York: HarperCollins, 2002), 283.
44 "The Greatest Defunct Web Sites and Dotcom Disasters," CNET, June 5, 2008, http://web.archive.org/web/20080607211840/http://crave.cnet.co.uk/0,39029477,49296926-6,00.htm.
45 Jim Edwards, "One of the Kings of the '90s Dot-com Bubble Now Faces 20 Years in Prison," *Business Insider*, December 6, 2016, http://www.businessinsider.com/where-are-the-kings-of-the-1990s-dot-com-bubble-bust-2016-12/#petscoms-greg-mclemore-raised-121-million-from-investors-but-lost-money-on-every-sale-7.
46 Cassidy, *Dot.Con*, 273.
47 Ibid., 306.
48 Ibid., 292.
49 David Kleinbard, "The $1.7 Trillion Dot.com Lesson," CNNMoney, November 9, 2000, http://cnnfn.cnn.com/2000/11/09/technology/overview/.
50 Zhu Wang, "Technological Innovation and Market Turbulence: The Dot-com Experience," *Review of Economic Dynamics* 10, no. 1 (2007): 78, 79.
51 Don Clark, "PayPal Files for an IPO, Testing a Frosty Market," *Wall Street Journal*, October 1, 2001.
52 Saul Hansell, "Some Hard Lessons for Online Grocer," *New York Times*, February 19, 2001.
53 Karen Angel, *Inside Yahoo!: Reinvention and the Road Ahead* (New York: John Wiley, 2002), 222.
54 Stephan Paternot, *A Very Public Offering: A Rebel's Story of Business Excess, Success, and Reckoning* (New York: John Wiley, 2001), 67.
55 "Silicon Alley 100," *Silicon Alley Reporter*, March 1999.
56 Paternot, *A Very Public Offering*, 111.
57 Cassidy, *Dot.Con*, 197.
58 Paternot, *A Very Public Offering*, 118.
59 Ibid., 201.
60 Lessley Anderson, "The Selling of TheGlobe.com," *Industry Standard*, July 5–12, 1999.
61 Securities and Exchange Commission, Form 10-Q, quarterly report for TheGlobe.com, accessed February 1, 2018, https://www.sec.gov/Archives/edgar/data/1066684/000089534500000280/0000895345-00-000280.txt.
62 Alan Abelson, "Up & Down Wall Street," *Barron's*, August 14, 2000.
63 David Henry, "More Insiders Sell Big Blocks of Stock: Surge May Foretell Market Weakness in 3 to 12 Months," *USA Today*, September 18, 2000, 1B.
64 James Altucher, "How I Helped Mark Cuban Make a Billion Dollars and 5 Things I Learned from Him," posted 2017, http://www.jamesaltucher.com/2011/04/why-im-jealous-of-mark-cuban-and-5-things-i-learned-from-him/.
65 Maggie Mahar, *Bull! A History of the Boom and Bust, 1982–2004* (New York: HarperBusiness, 2003), 319.
66 Casey Hait and Stephen Weiss, *Digital Hustlers: Living Large and Falling Hard in Silicon Alley* (New York: HarperCollins, 2001), 292.

67 Mahar, *Bull!*, 325.
68 Ibid., 333.
69 Chet Currier, "The Bear Market Is Dead—Long Live the New Bull," Bloomberg News, June 13, 2003.
70 "Participants Report Card for 2002: The Impact of the Bear Market on Retirement Savings Plans," Vanguard Group Retirement Research, February 2003.
71 John Markoff, "Why Google Is Peering Out, at Microsoft," *New York Times*, May 3, 2004.
72 *Tim Ferriss Show*, "163: Marc Andreessen—Lessons, Predictions, and Recommendations from an Icon," https://tim.blog/2016/05/29/marc-andreessen/.
73 Keith Collins and David Ingold, "Through Years of Tumult, AOL Sticks Around," Bloomberg, posted May 12, 2015, https://www.bloomberg.com/graphics/2015-verizon-aol-deal/.
74 Jim Hu, "AOL Time Warner Drops AOL from Name," CNET, September 18, 2003, https://www.cnet.com/news/aol-time-warner-drops-aol-from-name/.
75 Swisher, *There Must Be a Pony in Here Somewhere*, 220, 260.
76 Christian Wolmar, *Fire and Steam: A New History of the Railways in Britain* (London: Atlantic Books, 2007), locs. 1628, 1971–72, Kindle.
77 Ibid., loc. 1934–35.
78 Ibid., loc. 1941–44.
79 Ibid., loc. 1637–38.
80 Om Malik, *Broadbandits: Inside the $750 Billion Telecom Heist* (Hoboken, NJ: John Wiley, 2003), x.
81 Roger Lowenstein, *Origins of the Crash: The Great Bubble and Its Undoing* (New York: Penguin, 2004), 150.
82 Malik, *Broadbandits*, xi; Shawn Young, "Why the Glut in Fiber Lines Remains Huge," *Wall Street Journal*, May 12, 2005.
83 *Wired* Staff, "Bandwidth Glut Lives On," *Wired*, September 30, 2004, http://archive.wired.com/techbiz/media/news/2004/09/65121?currentPage=all.
84 Young, "Why the Glut in Fiber Lines Remains Huge."
85 "Internet Users in the World," Internet Live Stats, http://www.internetlivestats.com/internet-users/.
86 "Total Number of Websites," Internet Live Stats, http://www.internetlivestats.com/total-number-of-websites/.
87 Angel, *Inside Yahoo!*, 173.
88 Erick Schonfeld, "Facebook Overthrows Yahoo to Become the World's Third Largest Website," TechCrunch, December 24, 2010, https://techcrunch.com/2010/12/24/facebook-yahoo-third-largest-website/.
89 "Zuckerberg, Facebook Move to Mimic Amazon and Google's 'Go Anywhere' Strategy," *Peridot Capitalist*, April 17, 2014, https://www.peridotcapitalist.com/2014/04/.

11. I'M FEELING LUCKY: GOOGLE, NAPSTER AND THE REBIRTH

1 John Battelle, "The Birth of Google," *Wired*, August 1, 2005.
2 Steven Levy, *In the Plex: How Google Thinks, Works, and Shapes Our Lives* (New York: Simon & Schuster, 2011), 121–22.

3 David A. Vise, *The Google Story: For Google's 10th Birthday* (New York: Delta, 2006), 20.
4 Ibid., 37.
5 Levy, *In the Plex*, 17.
6 Ibid., 21.
7 Vise, *The Google Story*, 38.
8 Ibid., 33.
9 Levy, *In the Plex*, 29.
10 John Battelle, *The Search: How Google and Its Rivals Rewrote the Rules of Business and Transformed Our Culture* (New York: Portfolio, 2005), 84.
11 Internet History Podcast, *Episode 41: Excite.com CEO George Bell*, November 17, 2014.
12 Battelle, *The Search*, 85.
13 Levy, *In the Plex*, 31.
14 Vise, *The Google Story*, 79.
15 Battelle, *The Search*, 89.
16 Vise, *The Google Story*, 85.
17 Michael Specter, "Search and Deploy," *New Yorker*, May 29, 2000.
18 Vise, *The Google Story*, 96.
19 David Kirkpatrick, "What's a Google? A Great Search Engine, That's What," *Fortune*, November 8, 1999.
20 Levy, *In the Plex*, 72.
21 Ibid., 36.
22 Ibid., 67.
23 Steve O'Hear, "Inside the Billion-Dollar Hacker Club," TechCrunch, March 2, 2014, https://techcrunch.com/2014/03/02/w00w00/.
24 Internet History Podcast, *Episode 73: "Father" of the MP3, Karlheinz Brandenburg*, July 14, 2015.
25 Ibid.
26 "A History of Storage Cost," mkomo.com, September 8, 2009, http://www.mkomo.com/cost-per-gigabyte.
27 David Essex, "More Big Honkin' Hard Drives in 1999," CNN.com, January 21, 1999, http://www.cnn.com/TECH/computing/9901/21/honkin.idg/.
28 Paul Boutin, "Nullsoft, 1997–2004: AOL Kills Off the Last Maverick Tech Company," *Slate*, November 12, 2004, http://www.slate.com/articles/technology/webhead/2004/11/nullsoft_19972004.html.
29 *Downloaded*, documentary directed by Alex Winter, 2013.
30 Joseph Menn, *All the Rave: The Rise and Fall of Shawn Fanning's Napster* (New York: Crown Business, 2003), 191.
31 Ibid., 247, 260.
32 Ibid., 223.
33 Ibid., 134.
34 Richard Nieva, "Ashes to Ashes, Peer to Peer: An Oral History of Napster," *Fortune*, September 5, 2013.
35 Menn, *All the Rave*, 205.
36 Nieva, "Ashes to Ashes, Peer to Peer."
37 Internet History Podcast, *Episode 139: The Napster Story with Jordan Ritter*, April 16, 2017.

38 Greg Kot, *Ripped: How the Wired Generation Revolutionized Music* (New York: Scribner, 2010), 31.
39 Steve Knopper, *Appetite for Self-Destruction: The Spectacular Crash of the Record Industry in the Digital Age* (New York: Free Press, 2009),135.
40 Menn, *All the Rave*, 144.
41 Ibid., 230.
42 Ibid., 244.
43 Knopper, *Appetite for Self-Destruction*, 148.
44 *Downloaded.*
45 Menn, *All the Rave*, 102.
46 Knopper, *Appetite for Self-Destruction*, 143.
47 Kot, *Ripped*, 45.
48 *Downloaded.*
49 Stephen W. Webb, "*RIAA v. Diamond Multimedia Systems:* The Recording Industry Attempts to Slow the MP3 Revolution, Taking Aim at the Jogger Friendly Diamond Rio," *Richmond Journal of Law and Technology* 7, no. 1 (Fall 2000), at https://scholarship.richmond.edu/cgi/viewcontent.cgi?article=1102&context=jolt.
50 Stephen Witt, *How Music Got Free: The End of an Industry, the Turn of the Century, and the Patient Zero of Piracy* (New York: Viking, 2015), 126.

12. RIP. MIX. BURN.: THE IPOD, ITUNES, AND NETFLIX

1 Joe Wilcox, "Apple: Looking for a Few Good Converts," CNET, March 26, 2002.
2 Alyson Raletz, "Man Who Came Up with iMac Name Tells What the 'i' Stands For," *Kansas City Business Journal*, June 7, 2012.
3 Brent Schlender and Rick Tetzeli, *Becoming Steve Jobs: The Evolution of a Reckless Upstart into a Visionary Leader* (New York: Crown Business, 2016), p. 263, Kindle.
4 Walter Isaacson, *Steve Jobs* (New York: Simon & Schuster, 2011), 384, Kindle.
5 Steve Knopper, *Appetite for Self-Destruction: The Spectacular Crash of the Record Industry in the Digital Age* (New York: Free Press, 2009), 166.
6 Isaacson, *Steve Jobs*, 388, Kindle.
7 Ibid.
8 Steven Levy, *The Perfect Thing: How the iPod Shuffles Commerce, Culture, and Coolness* (New York: Simon & Schuster, 2006), 134–35, Kindle.
9 Leander Kahney, *Jony Ive: The Genius Behind Apple's Greatest Products* (New York: Portfolio, 2013), 183.
10 Isaacson, *Steve Jobs*, 390.
11 "Steve Jobs Introduces Original iPod—Apple Special Event (2001)," posted January 4, 2014, https://www.youtube.com/watch?v=SYMTy6fchiQ.
12 Schlender and Tetzeli, *Becoming Steve Jobs*, 272.
13 Kot, *Ripped*, 35.
14 Ibid., 43.
15 Ibid., 42.
16 Isaacson, *Steve Jobs*, 396.
17 Ibid.

18 Ibid., 403.
19 Levy, *The Perfect Thing*, 143.
20 Isaacson, *Steve Jobs*, 405.
21 Ibid.
22 Levy, *The Perfect Thing*, 105.
23 Ibid., 109.
24 Ibid., 58.
25 Knopper, *Appetite for Self-Destruction*, 232.
26 "Global Recorded Music Revenue from 2002 to 2016 (in Billion U.S. Dollars)," https://www.statista.com/statistics/272305/global-revenue-of-the-music-industry/.
27 Tweet from @Mark_J_Perry, sourced from the RIAA, April 18, 2017, https://twitter.com/mark_j_perry/status/854407708870660097.
28 Knopper, *Appetite for Self-Destruction*, 181.
29 Gina Keating, *Netflixed: The Epic Battle for America's Eyeballs* (New York: Portfolio, 2012), 27.
30 Stephen P. Kaufman and Willy Shih, *Netflix in 2011*, Harvard Business Review Case Study 615-007 (August 2014).
31 Keating, *Netflixed*, 67.
32 Ibid., 27.
33 Ibid., 59.
34 Ibid.
35 Peter J. Coughlan and Jennifer L. Illes, *Blockbuster Inc. & Technological Substitution (A): Achieving Dominance in the Video Rental Industry*, Harvard Business Review Case Study 9-704-404 (December 18, 2003).
36 Sunil Chopra and Murali Veeraiyan, *Movie Rental Business: Blockbuster, Netflix, and Redbox*, Harvard Business Review Case Study KEL616 (October 12, 2010).
37 Daniel Kadlec, "How Blockbuster Changed the Rules," *Time*, August 3, 1998.
38 Kaufman and Shih, *Netflix in 2011*.
39 Keating, *Netflixed*, 185.
40 Kaufman and Shih, *Netflix in 2011*; Chopra and Veeraiyan, *Movie Rental Business*.
41 Kaufman and Shih, *Netflix in 2011*.
42 Jeremy O'Brien, "The Netflix Effect," *Wired*, December 1, 2012, https://www.wired.com/2002/12/netflix-6/.
43 Ibid.
44 Larry Downes and Paul Nunes, "Blockbuster Becomes a Casualty of Big Bang Disruption," *Harvard Business Review*, November 3, 2013.
45 Coughlan and Illes, *Blockbuster Inc. & Technological Substitution (A)*.
46 Maria Halkias, "Blockbuster Is Trying to Turn It Around," *Dallas Morning News*, May 2010, https://www.dallasnews.com/business/business/2010/05/08/Blockbuster-is-trying-to-turn-it-3330.
47 Chopra and Veeraiyan, *Movie Rental Business*.
48 Coughlan and Illes, *Blockbuster Inc. & Technological Substitution (A)*.
49 Conor Knighton, "Be Kind, Rewind: Blockbuster Stores Kept Open in

Alaska," *CBS Sunday Morning*, April 23, 2017, https://www.cbsnews.com/news/be-kind-rewind-blockbuster-stores-kept-open-in-alaska/?ftag=CNM-00-10aab8c&linkId=36799161.

50 O'Brien, "The Netflix Effect."

51 Matthew Boyle, "Questions for . . . Reed Hastings," *Fortune*, May 23, 2007.

13. A THOUSAND FLOWERS, BLOOMING: PAYPAL, ADWORDS, GOOGLE'S IPO AND BLOGS

1 Fara Warner, "These Guys Will Make You Pay," *Fast Company*, November 2001.

2 Ibid.

3 Eric M. Jackson, *The PayPal Wars: Battles with eBay, the Media, the Mafia, and the Rest of Planet Earth* (Washington, DC: WND Books, 2012), 34, 40.

4 Ibid., 180–81.

5 Matt Richtel, "Internet Offering Soars, Just Like Old Times," *New York Times*, February 16, 2002.

6 John Battelle, *The Search: How Google and Its Rivals Rewrote the Rules of Business and Transformed Our Culture* (New York: Portfolio, 2005), 126.

7 David A. Vise, *The Google Story: For Google's 10th Birthday* (New York: Delta, 2006), 98.

8 Battelle, *The Search*, 123.

9 Ibid., 93.

10 Kevin J. Delaney, "After Google's IPO, Can Ads Keep Fueling Company's Engine?" *Wall Street Journal*, April 29, 2004.

11 Ben Elgin, Linda Himelstein, Ronald Grover, and Heather Green, "Inside Yahoo!," *BusinessWeek*, May 21, 2001.

12 Sergey Brin and Lawrence Page, "The Anatomy of a Large-Scale Hypertextual Web Search Engine," accessed February 1, 2018, http://infolab.stanford.edu/~backrub/google.html.

13 Jim Hu, "Yahoo Reports Profit on Higher Revenue," CNET, October 9, 2002.

14 Battelle, *The Search*, 141.

15 Steven Levy, *In the Plex: How Google Thinks, Works, and Shapes Our Lives* (New York: Simon & Schuster, 2011), 94.

16 Vise, *The Google Story*, 119.

17 Battelle, *The Search*, 148; Levy, *In the Plex*, 70.

18 "Google's Ad Revenue from 2001 to 2016," 2018, https://www.statista.com/statistics/266249/advertising-revenue-of-google/; John Huey, Martin Nisenholtz, and Paul Sagan, *Riptide* (Cambridge, MA: Harvard University/Shorenstein Center on Media, Politics and Public Policy, 2013), vol. 1, chap. 12, https://www.digitalriptide.org/chapter-12-google-the-second-coming/.

19 Kevin J. Delaney and Robin Sidel, "Google IPO Aims to Change the Rules," *Wall Street Journal*, April 30, 2004.

20 Levy, *In the Plex*, 150.

21 Battelle, *The Search*, 220

22 Delaney, "After Google's IPO."

23 Ibid.
24 Matt Richtel, "Analysts Doubt Public Offering of Google Is a Bellwether," *New York Times*, May 1, 2004.
25 "Letter from the Founders," *Wall Street Journal*, updated April 29, 2004, https://www.wsj.com/articles/SB108326432110097510.
26 "Excerpts from Google's Filing," *Wall Street Journal*, updated April 29, 2004, https://www.wsj.com/articles/SB108326291882697484.
27 Levy, *In the Plex*, 149.
28 Ibid., 151.
29 Ibid., 149.
30 Gregory Zuckerman, "Google Shares Prove Big Winners—for a Day," *Wall Street Journal*, August 20, 2004.
31 Ian Ayres and Barry Nalebuff, "Going, Going, Google," *Wall Street Journal*, August 20, 2004.
32 Battelle, *The Search*, 227.
33 Laurie J. Flynn, "The Google I.P.O.: The Founders; 2 Wild and Crazy Guys (Soon to Be Billionaires), and Hoping to Keep It That Way," *New York Times*, April 30, 2004.
34 John Markoff, "Why Google Is Peering Out, at Microsoft," *New York Times*, May 3, 2004.
35 Levy, *In the Plex*, 101, 102.
36 Scott Rosenberg, *Say Everything: How Blogging Began, What It's Becoming, and Why It Matters* (New York: Crown, 2009), 120, 125.
37 Ibid., 101.
38 Ibid., 102.
39 Ibid., 18.
40 Ibid., 53.
41 *Newsweek* Staff, "Whispers on the Web," *Newsweek*, August 17, 1997, http://www.newsweek.com/whispers-web-172450.
42 Matt Drudge, "Anyone with a Modem Can Report on the World," address before the National Press Club, June 2, 1998, http://www.bigeye.com/drudge.htm.
43 Ibid.
44 Philip Weiss, "Watching Matt Drudge," *New York*, August 24, 2007.
45 Brian Abrams, *Gawker: An Oral History* (Kindle Single, 2015), loc. 138.
46 Ibid., loc. 229–30.
47 Ibid., loc. 248.
48 Julie Bosman, "First with the Scoop, If Not the Truth," *New York Times*, April 18, 2004, http://www.nytimes.com/2004/04/18/style/first-with-the-scoop-if-not-the-truth.html?_r=0.
49 Vanessa Grigoriadis, "Everybody Sucks," *New York*, October 14, 2007.

14. WEB 2.0: WIKIPEDIA, YOUTUBE AND THE WISDOM OF CROWDS

1 Nick Denton, "Second Sight," *Guardian*, September 20, 2001.
2 Scott Rosenberg, *Say Everything: How Blogging Began, What It's Becoming, and Why It Matters* (New York: Crown, 2009), 38.

3 Sarah Lacy, *Once You're Lucky, Twice You're Good: The Rebirth of Silicon Valley and the Rise of Web 2.0* (New York: Gotham, 2008), 6.
4 "Jurisimprudence," *Schott's Vocab*, May 31, 2010, https://schott.blogs.nytimes.com/2010/05/31/jurisimprudence/.
5 Andrew Lih, *The Wikipedia Revolution: How a Bunch of Nobodies Created the World's Greatest Encyclopedia* (New York: Hyperion, 2009), xv.
6 Ibid., 64–65.
7 "User: Ben Kovitz," Wikipedia, last modified December 20, 2017, https://en.wikipedia.org/wiki/User:BenKovitz#The_conversation_at_the_taco_stand.
8 "Wikipedia Statistics: English," December 18, 2017, https://stats.wikimedia.org/EN/TablesWikipediaEN.htm.
9 "Web 2.0," November 2005, http://www.paulgraham.com/web20.html.
10 Fred Vogelstein, "TechCrunch Blogger Michael Arrington Can Generate Buzz . . . and Cash," *Wired*, June 22, 2007.
11 Julia Angwin, *Stealing MySpace: The Battle to Control the Most Popular Website in America* (New York: Random House, 2009), 59.
12 Ibid., 238.
13 National Venture Capital Association, *Yearbook 2015*, http://nvca.org/?ddownload=1868.
14 Associated Press, "Venture Investment Hits a 6-Year High," *Los Angeles Times*, January 19, 2008, http://articles.latimes.com/2008/jan/19/business/fi-venture19.
15 Lacy, *Once You're Lucky, Twice You're Good*, 100.
16 Sarah Lacy and Jessi Hempel, "Valley Boys," *BusinessWeek*, August 14, 2006.
17 Michael Arrington, "Digg Is (Almost) as Big as Slashdot," TechCrunch.com, November 9, 2005, https://techcrunch.com/2005/11/09/digg-is-almost-as-big-as-slashdot/.
18 Lacy and Hempel, "Valley Boys."
19 Lacy, *Once You're Lucky, Twice You're Good*, 76.
20 John Cloud, "The YouTube Gurus," *Time*, December 25, 2006.
21 Steven Levy, *In the Plex: How Google Thinks, Works, and Shapes Our Lives* (New York: Simon & Schuster, 2011), 245.
22 *Wired* Staff, "Now Starring on the Web: YouTube," *Wired*, April 9, 2006, http://archive.wired.com/techbiz/media/news/2006/04/70627.
23 Randall Stross, *Planet Google: One Company's Audacious Plan to Organize Everything We Know* (New York: Free Press, 2008), 193.
24 Jason Abbruzzese, "The Rise and Fall of AIM, the Breakthrough AOL Never Wanted," Mashable, April 15, 2014, http://mashable.com/2014/04/15/aim-history/#IJvEwv67sPq3.
25 Angwin, *Stealing MySpace*, 52.
26 "The Father of Social Networking," Mixergy, December 3, 2014, https://mixergy.com/interviews/andrew-weinreich-sixdegrees/.
27 David Kirkpatrick, *The Facebook Effect: The Inside Story of the Company That Is Connecting the World* (New York: Simon & Schuster, 2010), 69.
28 Angwin, *Stealing MySpace*, 53.

29 Gary Rivlin, "Wallflower at the Web Party," *New York Times*, October 15, 2006.
30 Internet History Podcast, *Episode 117: Founder of Friendster and Nuzzel, Jonathan Abrams*, September 18, 2016.
31 Angwin, *Stealing MySpace*, 64.
32 Lev Grossman, "Tila Tequila," *Time*, December 16, 2006.
33 Angwin, *Stealing MySpace*, 84, 103.
34 Ibid., 140.
35 Ibid., 104.
36 John Cassidy, "Me Media: How Hanging Out on the Internet Became Big Business," *New Yorker*, May 15, 2006.
37 Angwin, *Stealing MySpace*, 175, 179.
38 Ibid., 262.

15. THE SOCIAL NETWORK: FACEBOOK

1 S. F. Brickman, "Not-So-Artificial Intelligence," *Crimson*, October 23, 2003.
2 David Kirkpatrick, *The Facebook Effect: The Inside Story of the Company That Is Connecting the World* (New York: Simon & Schuster, 2010), 25.
3 Ben Mezrich, *The Accidental Billionaires: The Founding of Facebook* (New York: Anchor Books, 2010), 49.
4 Kirkpatrick, *The Facebook Effect*, 26.
5 *Crimson* Staff, "Put Online a Happy Face," *Crimson*, December 11, 2003.
6 Luke O'Brien, "Poking Facebook," *02138*, November–December 2007, 66.
7 Mezrich, *The Accidental Billionaires*, 95.
8 Sam Altman, "Mark Zuckerberg on How to Build the Future," *Y Combinator* (blog), August 16, 2016, http://blog.ycombinator.com/mark-zuckerberg-future-interview/.
9 Kirkpatrick, *The Facebook Effect*, 34.
10 Ibid., 38.
11 "CS50 Lecture by Mark Zuckerberg," December 7, 2005; posted April 4, 2014, https://www.youtube.com/watch?v=xFFs9UgOAlE.
12 Kirkpatrick, *The Facebook Effect*, 38.
13 Ibid., 47.
14 Katherine Losse, *The Boy Kings: A Journey into the Heart of the Social Network* (New York: Free Press, 2012), xvii.
15 O'Brien, "Poking Facebook."
16 Kirkpatrick, *The Facebook Effect*, 43.
17 Ibid., 42.
18 Cassidy, "Me Media."
19 Kirkpatrick, *The Facebook Effect*, 64.
20 Kevin J. Feeney, "Business, Casual," *Crimson*, February 24, 2005.
21 Ibid.
22 Sarah Lacy, *Once You're Lucky, Twice You're Good: The Rebirth of Silicon Valley and the Rise of Web 2.0* (New York: Gotham, 2008), 150.
23 Ibid.
24 Kirkpatrick, *The Facebook Effect*, 63.
25 Ibid. 48.

26 Ibid., 89.
27 Altman, "Mark Zuckerberg on How to Build the Future."
28 Kirkpatrick, *The Facebook Effect*, 86.
29 Ibid., 95.
30 Ibid., 103.
31 Ibid., 98.
32 "What's the Story Behind Mark Zuckerberg's Fabled 'I'm CEO . . . Bitch!' Business Card?" updated February 1, 2011, https://www.quora.com/Facebook-company/Whats-the-story-behind-Mark-Zuckerbergs-fabled-Im-CEO%E2%80%A6bitch-business-card/answer/Andrew-Boz-Bosworth.
33 Melia Robinson, "How Sean Parker Bounced Back from Being Fired to Change Facebook's History," *Business Insider*, February 9, 2015, http://www.businessinsider.com/how-plaxo-and-sean-parker-changed-facebook-2015-2.
34 Kirkpatrick, *The Facebook Effect*, 100.
35 Ibid.
36 Julia Angwin, *Stealing MySpace: The Battle to Control the Most Popular Website in America* (New York: Random House, 2009), 177.
37 "James W. Breyer and Mark E. Zuckerberg Interview, Oct. 26, 2005, Stanford University," posted July 14, 2012, https://www.youtube.com/watch?v=WA_ma359Meg&feature=youtu.be.
38 Kirkpatrick, *The Facebook Effect*, 149.
39 Ibid., 111.
40 "CS50 Lecture by Mark Zuckerberg."
41 Kirkpatrick, *The Facebook Effect*, 113.
42 Ibid., 126.
43 Ibid., 130.
44 Ibid., 148.
45 Ibid., 145.
46 Ibid., 131.
47 Ibid., 150.
48 Ibid., 152.
49 Ibid., 154.
50 "CS50 Lecture by Mark Zuckerberg."
51 Kirkpatrick, *The Facebook Effect*, 156.
52 Ibid., 157.
53 Ibid.
54 Ibid., 170.
55 Ibid., 168.
56 Allison Fass, "Peter Thiel Talks About the Day Mark Zuckerberg Turned Down Yahoo's $1 Billion," Inc.com, March 12, 2013, https://www.inc.com/allison-fass/peter-thiel-mark-zuckerberg-luck-day-facebook-turned-down-billion-dollars.html.
57 David Kushner, "The Baby Billionaires of Silicon Valley," *Rolling Stone*, November 16, 2006.
58 Kirkpatrick, *The Facebook Effect*, 161.
59 Ibid., 168.

60 Lacy, *Once You're Lucky, Twice You're Good*, 169.
61 Kirkpatrick, *The Facebook Effect*, 180.
62 Ibid., 181.
63 Ibid., 189.
64 Ibid., 190.
65 Ibid.
66 Ibid., 192.
67 Ibid., 191.
68 Ibid., 192.
69 Ibid., 173.
70 Ibid., 185.
71 Ibid., 197.
72 Ibid., 227.
73 Ellen McGirt, "Hacker. Dropout. CEO.," *Fast Company*, May 2007.
74 Kirkpatrick, *The Facebook Effect*, 235.
75 Ibid., 275.

16. THE RISE OF MOBILE: PALM, BLACKBERRY AND SMARTPHONES

1 Tom Hormby, "The Story Behind Apple's Newton," *Gizmodo*, January 19, 2010, https://gizmodo.com/5452193/the-story-behind-apples-newton.
2 Markos Kounalakis, *Defying Gravity: The Making of Newton* (Hillsboro, OR: Beyond Words, 1993), 01:56.
3 Andrea Butter and David Pogue, *Piloting Palm: The Inside Story of Palm, Handspring, and the Birth of the Billion-Dollar Handheld Industry* (New York: Wiley, 2002), 23.
4 "Newton Message Pad," apple-history, last modified July 15, 2015, http://apple-history.com/nmp.
5 Kounalakis, *Defying Gravity*, 00:36.
6 Jim Louderback, "Newton's Capabilities Just Don't Measure Up," *PC Week*, September 13, 1993.
7 Peter H. Lewis, "So Far, the Newton Experience Is Less Than Fulfilling," *New York Times*, September 26, 1993.
8 Harry McCracken, "Newton, Reconsidered," *Time*, June 1, 2002; John Markoff, "Apple's Newton Reborn: Will It Still the Critics?" *New York Times*, March 4, 1994.
9 David S. Evans, *Invisible Engines: How Software Platforms Drive Innovation and Transform Industries* (Cambridge, MA: MIT Press, 2006), 159.
10 Butter and Pogue, *Piloting Palm*, 197.
11 Evans, *Invisible Engines*, 155; Butter and Pogue, *Piloting Palm*, 166.
12 Rod McQueen, *BlackBerry: The Inside Story of Research in Motion* (Toronto, ON: Key Porter, 2010), 154.
13 Alastair Sweeny, *BlackBerry Planet: The Story of Research in Motion and the Little Device That Took the World by Storm* (Mississauga, ON: John Wiley, 2009), 47.
14 McQueen, *BlackBerry*, 93.
15 Ibid., 174.

16 Ibid., 185.
17 Sweeny, *BlackBerry Planet*, 4.
18 Jacquie McNish and Sean Silcoff, *Losing the Signal: The Untold Story Behind the Extraordinary Rise and Spectacular Fall of BlackBerry* (New York: Flatiron Books, 2015), 112.
19 Kevin Maney, "BlackBerry: The Heroin of Mobile Computing," *USA Today*, May 7, 2001.
20 McQueen, *BlackBerry*, 194.
21 Ibid., 80.
22 Brian Merchant, *The One Device: The Secret History of the iPhone* (New York: Little, Brown, 2017), 30.
23 Ibid., 34.
24 Ibid., 195.
25 Mary Meeker, Scott Devitt, and Liang Wu, *Internet Trends: June 7, 2010, CM Summit—New York City* (Morgan Stanley, 2010), www.kpcb.com/file/june-2010-internet-trends.
26 McQueen, *BlackBerry*, 11.
27 "Steve Jobs on Apple's Resurgence: 'Not a One-Man Show,'" BusinessWeek Online, May 12, 1998, available from the Internet Archive Wayback Machine, https://web.archive.org/web/20111209185106/http://www.businessweek.com/bwdaily/dnflash/may1998/nf80512d.htm.

17. ONE MORE THING: THE IPHONE

1 *How the iPhone Was Born: Inside Stories of Missteps and Triumphs*, video documentary, Wall Street Journal Video, June 25, 2017, http://www.wsj.com/video/how-the-iphone-was-born-inside-stories-of-missteps-and-triumphs/302CFE23-392D-4020-B1BD-B4B9CEF7D9A8.html.
2 Brian Merchant, *The One Device: The Secret History of the iPhone* (New York: Little, Brown, 2017), 200.
3 "Steve Jobs at D2 2004 All Things Digital Conference," March 25, 2013, https://www.youtube.com/watch?v=mCBu50CozH0.
4 "Steve Jobs in 2005 at D3," June 1, 2012, https://www.youtube.com/watch?v=IzH54FpWAP0.
5 *How the iPhone Was Born*.
6 Fred Vogelstein, *Dogfight: How Apple and Google Went to War and Started a Revolution* (New York: Sarah Crichton Books / Farrar, Straus and Giroux, 2013), p. 25, Kindle.
7 "Motorola Razr," note 3, Wikipedia, last modified December 26, 2017, https://en.wikipedia.org/wiki/Motorola_Razr#cite_note-3.
8 Charles Arthur, *Digital Wars: Apple, Google, Microsoft and the Battle for the Internet* (Philadelphia: Kogan, 2012), loc. 153, Kindle.
9 Frank Rose, "Battle for the Soul of the MP3 Phone," *Wired*, November 1, 2005.
10 Merchant, *The One Device*, 217.
11 Vogelstein, *Dogfight*, 28.
12 Ibid.
13 Ibid., 29.

14 Merchant, *The One Device*, 217.
15 Walter Isaacson, *Steve Jobs* (New York: Simon & Schuster, 2011), p. 466, Kindle.
16 Merchant, *The One Device*, 205.
17 Ibid., 20.
18 Ibid., 21.
19 Isaacson, *Steve Jobs*, 468.
20 Brent Schlender and Rick Tetzeli, *Becoming Steve Jobs: The Evolution of a Reckless Upstart into a Visionary Leader* (New York: Crown Business, 2016), p. 310, Kindle.
21 Merchant, *The One Device*, 94.
22 "CHM Live: Original iPhone Software Team Leader Scott Forstall (Part Two), June 28," 2017, https://www.youtube.com/watch?v=IiuVggWNqSA.
23 Merchant, *The One Device*, 105.
24 Vogelstein, *Dogfight*, 38.
25 Matthew Panzarino, "Apple v. Samsung Day 2: Schiller, Forstall Testify on Creation, Sales and Hardships of iPhone Project," *Next Web*, August 3, 2012, https://thenextweb.com/apple/2012/08/03/apple-v-samsung-day-2-schiller-forstall-testify-on-creation-sales-and-hardships-of-iphone-project/.
26 Merchant, *The One Device*, 209.
27 "On the Verge—Tony Fadell and Chris Grant—On the Verge, Episode 005," April 30, 2012, https://www.youtube.com/watch?v=qf9XcNWRvSU&t=1901s.
28 Isaacson, *Steve Jobs*, 469.
29 Ibid., 67.
30 Ibid., 35–36.
31 Merchant, *The One Device*, 365.
32 Vogelstein, *Dogfight*, 17.
33 Schlender and Tetzeli, *Becoming Steve Jobs*, 360.
34 "Steve Jobs Talks iPhone—All Things D5 (2007)," posted December 22, 2013, https://www.youtube.com/watch?v=fkPN_U0D3CM&t=1570s.
35 John Markoff, "Phone Shows Apple's Impact on Consumer Products," *New York Times*, January 11, 2007.
36 Schlender and Tetzeli, *Becoming Steve Jobs*, 363.
37 Ibid.
38 Ibid., 362.
39 Prince McLean, "Apple iPhone 3G Sales Surpass RIM's BlackBerry," AppleInsider, October 21, 2008, http://appleinsider.com/articles/08/10/21/apple_iphone_3g_sales_surpass_rims_blackberry.html.
40 Adam Lella, "U.S. Smartphone Penetration Surpassed 80 Percent in 2016," comScore, February 3, 2017, https://www.comscore.com/Insights/Blog/US-Smartphone-Penetration-Surpassed-80-Percent-in-2016.
41 "Steve Jobs Talks iPhone—All Things D5 (2007)."

INDEX

ABOUT THE AUTHOR

BRIAN MCCULLOUGH is an eighteen-year veteran of the Internet industry and the founder of various web-based startups. Host of the Techmeme Ride Home podcast and the Internet History Podcast, he was named a 2016 TED Resident. He lives in Brooklyn, New York.